The
Unified Principle
of Colour

The Unified Principle of Colour

PETER MODDEL

IBEX PRESS

The Unified Principle of Colour – 1st edition
ISBN Paperback: 978-2-9700967-2-6
ISBN eBook: 978-2-9700967-3-3

Cover and logo design: Geneviève Romang
Interior Design: Creative Publishing Book Design

Contents

PART I

LIST OF FIGURES

Acknowledgements

I wish to thank Dr. Bernard Jenni, Switzerland, who solved uncountable technical problems that arose in preparing this work. His resourceful assistance, already in the 1980s when I prepared my early experiments in colour, set the path for much of this work. I thank also Dr. Tony Partridge who took the time to read the first draft of this book and offered valuable advice and encouragement. In developing the background for this book on several occasions and for a period of months, I was offered a place to stay where I could reflect and write at my ease. In particular, I express my thanks for the kindness and generosity of Jakob and Felizitas in Dharamkot, near Dharamsala, India; and to Aleck, whose memory is with me; and to his family on the wonderful Carrowgarry farm in Co. Sligo, Ireland.

I thank my brother and the many friends who encouraged me throughout the writing of this book. I dedicate this book to the memory of Sive, my sister. She is ever an inspiration in my activities.

Introduction

I.1 The challenge

Colours are very much present in our experience of the world and yet, what they are remains an unsolved mystery. An evergrowing number of publications, particularly since the second half of last century, explore colour from the point of view of physics, philosophy, the neurosciences, anthropology, linguistics, and more. Even so, the process that gives rise to colour has escaped detection. There are 'almost' explanations, but none are complete. In Newton's time it seemed that physics would reveal the source of colour. Today the expectation of an answer has passed to the neurosciences. The cascade of new facts about neurobiological processes and cellular interactions of uncanny complexity relating to colour production are more than wondrous and yet, with all this, an unexplained gap in the process persists. I dare say, the question that Newton wished to answer is no less a question today than it was in his time.

Is there another approach to the question, what is colour? I suggest we extricate the issue momentarily from the thicket of research findings on vision and colour and view the question afresh from a more encompassing position. It then becomes apparent that a full description of colour and colour vision requires a third discipline, beyond physics and biology. The needed discipline entails the consciousness of the observer, without which there can be no perceiver and no perception.

Although this third discipline is spoken of as *consciousness* and also as *consciousness studies*, what is actually implied is subjectivity; that of a perceiver conscious of what is being looked at. The trio of *physics, biology* and the *subject's consciousness* together produce the colours we experience. There is no need for a hidden code in the brain that orchestrates colour vision. No place remains for the assumption that colours are hidden in white light or anchored to specific wavelengths. Instead, a simple description of the unified principle of colour emerges to describe what colour is, how it is produced, and why it forms. The chapters that follow describe this in full detail.

I.2 Two analogies

To illustrate that the subjectivity of the perceiver can be instrumental in determining what is seen, I propose two analogies. In the first of these, colour vision can be compared with the appearance of water in a mirage, where the observer makes sense of certain visual impressions by recognising what is seen as water. The unaccounted for mind activity of the individual subject is brought to bear and the result is that the perceiver sees water.[1] A second analogy to seeing colour is the impression of depth on a flat, two dimensional drawing. The observer introduces a sense of depth and with this makes sense of what is seen. To do so requires working out relationships of size, of angles, and more. Without being conscious of this personal involvement that enables the result, the perceiver sees a three-dimensional world.

These two experiences emerge from specific visual cues. Even so, it takes a living subject to resolve the conflicting visual input and,

---

[1] I chose a mirage to dramatise the fact that the perceiving subject, not the observed object, brings the understanding of what is seen.

without the need for conscious intervention, to achieve perception. The observer sees water and senses depth. Colour vision is a further example of that phenomenon: a perceptive act rooted in the perceiver's ability to reach an understanding. Here too, specific physical and biological conditions enable and guide the act of colour perception; however, the conditions themselves do not produce the result. Colour is only seen when an intelligent being creates a response that overcomes a specific impediment to seeing. The following chapters show how each colour serves to resolve a somewhat different visual challenge.

To admit that the intelligence of an observer is required for colour production opens up a perspective that is not described by neurophysiological, matter-based models. It introduces the function of mind.[2] An individual, conscious, living being with the capacity to understand engenders observable outcomes, such as the presence of colour.

1.3 The intent to see

In this book, colour is not assumed to be a feature inherent to light, nor an algorithm hidden in the brain. Instead, colour is shown to stem from an act of volition that can be understood as the intent to perceive a visible object. In the early chapters, I describe how, facing a visual impediment, the observer's steadfast desire to see calls forth a solution and generates the novel and surprising solution that we know as colour.

A frog fell into a bucket of cream (poor fellow) and, with unrelenting determination to escape, kept flailing about. After some time, the situation completely changed for, wonder of wonders, didn't the victorious frog find itself atop a motte of butter!

---

[2] It is mind, not brain (see Chapter 2 for a definition of mind).

With perseverance, the initial impossibility creates a new solution. Yes, I am proposing that attitude has an operational role! Faced by an unresolvable visual conundrum, the observer sustains the intent to see, the intent to render intelligible what meets the eye. It is as if determination calls for an answer and the universe conjures up a solution. This book describes the process in concrete terms. I show how the differing sensitivities of the retinal cells create an unsurmountable impasse that is overcome by the transformation of the colourless into something new. Exactly why and exactly how tones of grey transmute into colour is the central theme.

1.4 A change of emphasis

This approach introduces a fundamental reorientation of our viewpoint on colour perception summarised briefly below:

1. Colours are produced out of necessity and at the moment in which they are perceived.
2. The transition from seeing brightness to seeing colour requires an intelligent act on the part of the observer – a living being.
3. Resolution of the visual conflict and the achievement of perception is driven by a specific desire present in the observer, which can be understood as the intent to see.
4. Such an intended outcome suggests purpose and therefore the activity of perception lies outside the framework of physical causality and includes processes that are not described in classical science.

More generally, the phenomenon of colour vision is just one example of the manner in which everything we observe results from an inherent desire, or intent, to make sense of our sensory input. In the book *Making Sense,*[3] I describe the process that generates

---

[3] Peter Moddel, *Making Sense* (Fribourg, Ibex Press, 2014).

consciousness and perception while, here, in this book, I include that which relates directly to colour perception.

I.5 A note on my approach

The description of colour formation proposed in this work is an argument based on what we observe. It resolves some difficulties in contemporary colour theory and creates new questions. I should alert the reader to the fact that the approach is not built on new experimental work, but has been arrived at through a process of deduction based entirely on existing research, that is to say, on what is already known and published concerning colour phenomena.

I.6 The book's structure

Each chapter opens with an abstract as a quick reference of its content to allow the reader to pinpoint precisely what interests them. I have tried to make each chapter a self-contained unit, which does necessitate a certain degree of repetition, although this has been kept to a minimum.

The thirteen chapters of Part I present the essential features of the unified principle of colour as it manifests in subtractive colours. The epilogue and three appendices emphasise the necessary presence of a living subject.

The three chapters of Part II consider how colours arise on a black and white spinning top, known as the Benham top. It is included as a work in progress.

PART I

1
The Case of Colour Vision

ABSTRACT 1

Two facts

- An object that reflects light from only part of the full solar spectrum appears coloured.
- An object that reflects the full spectrum of light appears colourless, that is, shades of grey.

The argument

- Each cone set forms a separate black and white view of what is seen
- An internal brightness scale is formed by each cone set that inter-relates the brightness levels of what it sees
- When the light that reaches our eyes is not full spectrum, different cone sets will register contradictory brightness ratios
- A conflict in brightness thus manifests when the views of the different cones encounter one another
- The subject's intent to see calls for the differences to be resolved
- Since the scales of brightness were formed independently by each cone set, they will need to establish a common ground for comparison
- This impediment to their interaction creates an impasse in the visual process, because the three views need to integrate into a single view that can enter consciousness.

1.1 Basics

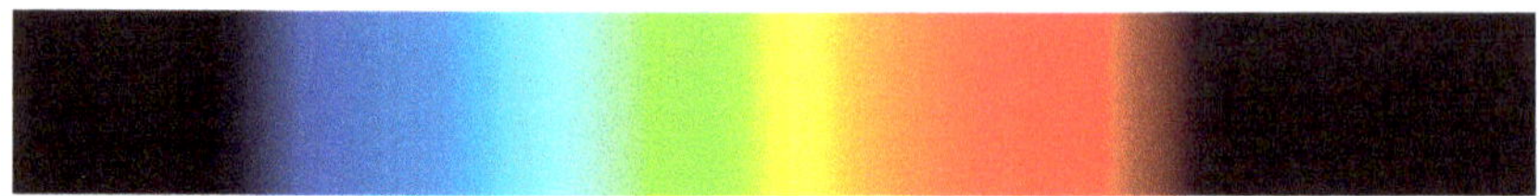

Figure 1. The spectrum of colours produced by a prism

An elementary question

What happens when a prism refracts a beam of sunlight? Surely everyone knows the answer – there's a rainbow of colour! But don't dismiss the question too quickly. If we state this obvious fact differently, it can lead our thinking in a new direction.

The obvious fact

The colours that exit from a prism spread out and the different wavelengths of light are separated. From this it is evident that each colour we see is a partial view of the full range of wavelengths.

The obvious conclusion

From this we realise that when we see only part of the full spectrum of sunlight, we see colour; the complement of this fact is that when we see the entire spectrum we see, not colour, but colourless illumination (black and white).

The function of colour

The absence of colour when the eye receives full-spectrum light suggests that colour is not required in that situation. The need for colour results only when light is not full spectrum. This is because colour has a specific function: it serves to rectify a visual imbalance or, more exactly, to resolve a contradiction that manifests when the light reaching our eyes is not full-spectrum sunlight.

1.2 The predicament

Let me begin by briefly setting out the wellknown, central facts of colour vision.

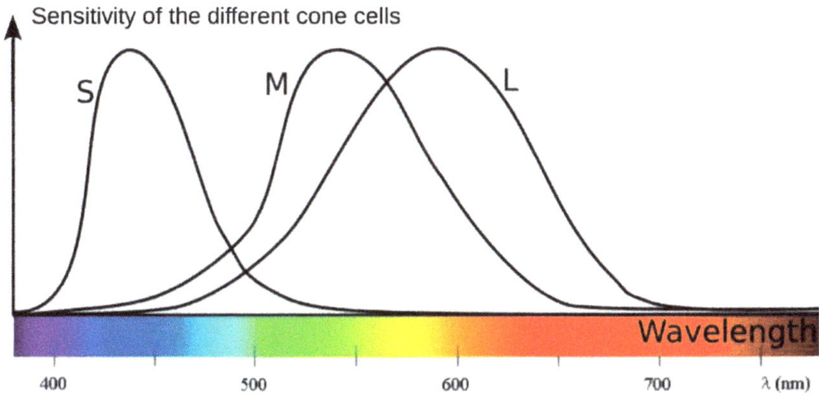

Figure 2. A standard diagram showing the wavelength sensitivities of the S, M, and L cones. With the spectrum colours underneath, this might suggest that wavelengths of light have colour. Not so – all wavelengths of light are colourless. Colour is produced in an entirely separate process.

Light coming through the iris of the eye falls upon the retina's light sensitive cells, called rods and cones. The cone cells are spread throughout the retina, but are more densely clustered in the fovea, which is the central area of the retina that receives light entering from the centre of the field of vision. Rod cells are absent in the fovea and, we're told, function in light that is too feeble to activate the cone cells.

Each cone cell is sensitive to one of three ranges of wavelengths within the visual spectrum:

- L cones: sensitive to a broad band, centred on the longer wavelengths
- M cones: sensitive to a less broad band, centred on medium wavelengths

- S cones: sensitive to a restricted band, centred on shorter wavelengths.

These three different ranges of sensitivity in colour vision were first deduced from the way colours appear. This was realised long before the 20th century and before microscopic analysis evidenced the presence of three different cones.[4]

Each cone set registers the brightness, or luminance, of the light in the wavelengths to which it is attuned. Cones of the same type generate the same image and produce one particular rendition of the scene being viewed. Since there are three different conetypes, in certain situations they can produce three incompatible views that contradict one another in brightness values.

It is important to understand that cones are sensitive to light intensity and not to colour. The display they register, were it visible, would be seen as a black and white image in shades of grey. But we are not conscious of the three black and white displays formed by each set of cones.

The primary reason these three distinct images do not enter our conscious mind is because of the mismatch, because the different views clash (as explained in the next section). Each preconscious view registers independently.[5] They do not fuse and, as will be shown, it is impossible for them to do so. In an advanced stage of the visual process, a transformation must take place that brings together the three colourless renditions of what is seen so they become a single

---

[4] The trichromaticity of human vision was discovered by two Londoners who never met each other, George Palmer and John Elliot in several publications issued between 1777 and 1796. It was later clearly formulated by Thomas Young in 1801'. V. Barsan, A. Merticariu, *Cogent Arts & Humanities* 3, no. 1 (2016): 1145569.

[5] A number of research findings support the fact that the displays formed by the three cone sets register on our visual system as three independent views. For example, Edwin H. Land, 'The Retinex Theory of Colour Vision,' Scientific American 237, no. 6 (December 1977): 108-28, 10.1038/scientificamerican1277-108.

intelligible image. Only after a unified view is formed do we become conscious of the presence of colour and, with this, we become conscious of what we are looking at.

1.3 Incongruence generated by the different cone sets

In the description that follows, the reader needs to bear in mind that cones only record brightness (luminance) and any scene they portray is always in black and white.

The brightness of objects in the field of view is related not only to the light that is present, but it also depends on the sensitivity of the cells. When the light coming from what we see contains only a partial spread of wavelengths, necessarily then the different cone sets will record the scene's brightness relationships differently, due to their separate wavelength sensitivities.

Let us consider what happens when we look at a hillside covered in coloured flowers. As perceivers, we'll speak of blue gentians and pink rhododendrons. But remember, each cone set is only registering brightness. To the L cones of the retina, the gentians' shorter wavelengths will appear less illuminated than the rhododendrons' longer wavelengths. For the S cones, being more sensitive to short wavelengths, the gentian flowers will be brighter, which means a grey closer to white, than the darker grey of the rhododendrons. The M cones, likewise, will produce a result relative to their sensitivity range.

In this way, the intensity (light grey or dark grey) of the view produced by each cone set will differ. But as to the form of what is seen and its spatial position, the three views will concur.[6] To the extent

____

[6] In the case of depth perception, when we view a scene with binocular vision, there is a parallel conflict, not in brightness but in spatial positions. This analogy is discussed in Chapter 3.

that the information from each cone set remains independent (more of this in the next section), the perceiver is confronted with three incompatible views. Without a unified view, a coherent image cannot form and, if anything was seen, three different views would pop in and out of consciousness.

1.4 The deterrent preventing the assimilation of the three views

Surely there is an obvious solution to the incompatible views? Combine them! For example, the brightness produced by the three separate daffodil renditions, each a different intensity of grey, could be added together. This blending of brightness levels would function for every object in the scene. Objects would then end up with the combined brightness of the three renditions and nothing would prevent the formation of a single black and white image of the visible scene.

This suggestion leads to the crux of the colour formation process. Such an assimilation of luminosities is impossible – and I will describe this impasse in detail.

To imagine that levels of brightness can be added together is to misunderstand how the level of brightness is established. Brightness is not read off an objective scale. There is no unit of brightness that could be employed as the 'standard candle' against which all shades of grey can be calibrated. I offer an analogy. In an orchestra tuning up before a concert, the oboe sounds out an A as the reference note, or frequency, to which a common harmonic structure can be anchored – and to this the other musicians adjust their instruments. However, in vision there is no such thing as a reference note that sets a common unit of brightness between the cone sets. Instead, a relative scale of brightness has to be generated spontaneously from the display that is being registered by the eye. It is as if the orchestral instruments found a common accord by playing together!

1.5 Brightness relationships

Figure 3. The Adelson checkerboard: On the left, the surfaces A and B appear as different. On the right we see that both are the same shade as the parallel grey lines.

The brightness we observe is not fully determined by the light that reaches our eyes from the object. Numerous experiences support the fact that brightness levels are relative. For example, a sheet of paper can appear white when it is placed between darker sheets, but hold it next to sunlit snow and it can appear a distinct grey. The luminosities present across the whole field of view influence how we see any single area, both in brightness and, as is well-known, in colour.

Relationships between luminosities are relative and form separately and internally within each cone set. Different scales of brightness form three separate views. However, sight can only result with the joint activity of all three cone sets, and this requires concord between their brightness relationships. The luminosity scales established separately by the three cone sets need to be brought together and resolved into a single unified view. This integration is achieved out of the range of the conscious mind.

Figure 4. A colour is influenced by the surrounding colours. The blue-green stripe seen between orange stripes tends to green, while between pink stripes, it tends to blue. However, the wide blue-green stripe bridging these examples shows that without surrounding colours, the difference disappears.

1.6 Reaching a common scale of reference

To return to the musical analogy, but with a choir this time, because singers have a greater ability to adjust to one another. They can be deeply attentive to each other and, while singing, find a common reference and adopt, for example, a particular key. This tonal centre can then function beautifully, but only as long as that choir sings alone. If they are joined by another choir (or two, as an analogy for the cone sets), a serious difficulty can arise if the other choir adopted an independent tonal centre in the same manner. The choirs would not have a common reference, even if they are singing the same piece of music. In the resultant cacophony, the listener would be unable to make intelligent sense of the whole.[7]

---

[7] Actually, for music, because of the harmonic structure we are familiar with (an octave of twelve semitones), the range of variability is smaller than with brightness, which lacks such a structured relationship.

The choir example is only helpful up to a point. Whereas we can hear the cacophony of sounds in the auditory case, in the visual case the three cone sets present three contradictory truths of what is before us. The result is a visual impasse and only when that is resolved do we become conscious of an image. The scale of brightness that functions for one cone set is isolated and separate from the scale established for another. The internal scale of brightness established by, say, the M cones, has no basis for comparison with the brightness registered by, say, the L cones. The luminosity values only make sense within a cone set's own, internally established scale. Comparisons between different cone views are not straightforward and can be compared with attempting to calculate with numbers that have different bases. They do not mix. A 2 in base 10 and a 2 in base 16 have different value scales and cannot be measured against one another. This fact is crucial in understanding the phenomenon of colour. Before going into this further, I offer an anecdote.

1.7 A curious thought!

Just imagine you were a chameleon, with one eye looking in one direction and the other eye in a different direction. This would be an insurmountable difficulty for a human, since we know ourselves to be a single observer, observing one scene at time. Simultaneously perceiving two or, in the case of colour, three different realities, bypasses our natural way of making visual sense of what we see.[8] Such is the contradiction that arises when light intensities do not match and three separate views vie with one another to be accepted as the true rendition of what is being looked at.

---

[8] Further implications are in Appendix I: The chameleon and our sense of self.

1.8 The path to follow

When looking at a scene not made up of full-spectrum light, three independent views that do not fit together are generated. Contradictions in the brightness relationships frustrates the ability to see as the perceiver is unable to form a single coherent view out of the visual input. Imagine the non-conscious mind pulsating with competing *potential* views, each preventing the others from entering consciousness.

To be sure, this whole battle only lasts a split second and the drama passes by unnoticed. But we need not be surprised by all the activity of which we are oblivious, for it takes place out of consciousness, where problems and solutions cohabit. In the study of colour, recognising both the impasse and the desire for resolution directs us to the trail that we will follow. We will trace the intermediary stages and, in Chapter 8, show how perceiver and perceived, in tandem, create the exquisite solution we know as colour.

Before setting out on this path, it is essential to have a clear understanding of the word *consciousness*. This is presented succinctly in Chapter 2.

2

Conscious Mind and Non-conscious Mind

ABSTRACT 2

- We use the word *consciousness* to refer to two distinct, contrasting modes of sensing the world and ourselves.

- Mode one is consciousness of an object. We are conscious of a particular meaning, that is to say, conscious of *something*. This is the common usage of the word consciousness. To be more precise and to clarify how it differs from the second mode, it can be called *conceptual consciousness*.

- Mode two relates to information received by a living being that does not enter conceptual consciousness. We can sense our surroundings and respond appropriately without consciously noting what is there – for instance, by stepping over a stone in our path without thinking about it. I refer to this subliminal form of integrating information exclusively as *expanded consciousness*.

- These two terms – *conceptual* and *expanded* – can also to refer to mind activity, with the *conceptual mind* and the *expanded mind*.

- The *conceptual mind* refers to mind activity that has specific content in thoughts and images.

- The *expanded mind* is the functioning of our intelligence beyond the reach of conceptual consciousness. For example: being faced with a question and, without thinking about it, a solution comes to mind, maybe on waking after sleep. The expanded mind, in a non-apparent process, puts an answer in our conscious mind.
- The conceptual mind functions in a linear manner, as thoughts and images succeed one another.
- When the mind moves beyond this linear constraint and diverges into multiple simultaneous activities, conceptual consciousness cannot follow. Quite naturally, the conscious mind, unable to enter multiple consciousness, imagines it to be the unconscious. In this way, the activity of expanded consciousness is overlooked and frequently not taken into consideration.

In describing the process of colour perception, the role of the perceiver needs to be taken into account and it is essential that the terms employed are clearly defined. Our first item is the word *consciousness*, especially because it has two divergent meanings that, when unrecognised, can lead to unnecessary argument. There are two distinct modes of consciousness.

2.1 Conceptual Consciousness versus Expanded Consciousness

Mode 1. Conceptual Consciousness, or simply, Consciousness

When we think of something, we are conscious of it. This, I refer to as *consciousness*. To differentiate it from the second mode (below), I refer to it also as *conceptual consciousness*. Both terms refer to the same thing and signify the recognition of an object, of something that occupies our thought. What we think of can be physical, imagined, or a feeling. With conceptual consciousness we acknowledge an object of consciousness.

Mode 2. Expanded Consciousness

A different mode of consciousness includes all the functioning mind deals with that does not register in our consciousness. Examples of this are, gathering the words that form a sentence while we are speaking; or we might be oblivious of sounds from a nearby stream or of bird calls in the branches above, while their effect, calming perhaps, influences our being. We can be aware of our surroundings and register their effect subliminally and respond appropriately without becoming conceptually conscious of the influence. This mode of consciousness that functions subliminally, I call *expanded consciousness*. Exclusively, this term is used to refer to it and not the word *consciousness* on its own.

2.2 The Conceptual Mind contrasted with the Expanded Mind

When speaking of the mind, it is necessary to discriminate between these same two distinct functions:

1. The first function is that of the mind that has manifest content. All that enters conceptual consciousness manifests as what we perceive and as the thoughts we think. I refer to this as the *conscious mind*, or with more precision, the *conceptually conscious mind*.

2. Beyond this is a different function of mind. It has the capacity to deal with information and integrate it into our being without our becoming conscious of receiving the information. That is, we are not conceptually conscious of it. This includes all we respond to that does not register in our conscious mind. For example, a person to whom we are speaking smiles. We do not notice the smile and yet we feel relaxed and well. Although the smile did not enter our conscious mind, on a non-conscious level it did enter and we responded to it and this implies that

on another level of consciousness, it was received by our living system. This subliminal function I refer to as the *expanded mind*.

2.3 Mind

With the word *mind*, I refer to the power of discernment that enables the integration of information into our living system. We receive information through diverse channels such as the intellect, the sensory system, and intuition.

To add further precision to this definition, with the word *mind*, we refer to the capacity to produce meaning. When meaning is absent, mind is absent. Stating it thus brings together the terms *mind* and *meaning*, two concepts that have been sidelined in scientific discourse. To clarify further my position, I add a footnote.[9]

---

[9] There is an intelligence to the body that knows how to respond when placed in a particular situation. The response can be an action or a state of being that the subject assumes, and it can manifest without registering in consciousness. It is the intelligent response of a living being to effects that were integrated into its experience. To make this more obvious, I add three examples.

First: To become upset requires the integration of all that forms that experience. The troubled person may not be conscious of sensing upset, but the mind has brought together the qualities that form that sensation. This is achieved through the activity of the expanded mind that integrates many diverse effects and transforms them into the meaning and sensation we, otherwise, verbalise as *upset*.

Second: A person responds to a melody that they did not notice was playing. Although not consciously noted, it generates a particular effect, maybe awakening a sense of pleasure or of nostalgia. This is possible because the living subject subliminally integrates the disparate sounds into that melody. This is achieved through the intelligence of the being hearing the sounds. The melody has become a new presence that no longer functions in the way disparate sounds do, for, as melody, they carry new meaning and a new impression. The capacity of integrating sounds so they become a particular expression is an ability of the expanded mind.

Third: A unicellular organism that withdraws from a disturbance or approaches a nutritive medium expresses a similar capacity. That behaviour is the expression of intelligence and reflects the expanded mind activity of that being.

With these examples, I call attention to the meaningful integration performed by a living being. Described in these terms, we can recognise mind as being an ever-present potential associated with a living being.

2.4 Awareness

The overall function of mind that includes the functions of both the conceptual mind and the expanded mind is what we refer to as *awareness*.

2.5 A linguistic precision

In English, the concept of *mind* has a subjective quality. Mind is the potential for intelligence and is attributed to a particular entity, a living being or a whole system. The relationship to a subject that animates or anchors the mind's content is implicit in the concept of mind. This key feature is lost in some translations. For example, in French the word mind is translated as *esprit* (spirit), which recasts its role. It catches the non-physical aspect of mind but loses the notion of the relationship between mind and the entity that sustains it. *Esprit* no longer signals the intelligence of a particular entity. In translation, mind as *esprit* has become objective, self-contained and purged of the sense of 'belonging to'. For the English speaker, mind and spirit are very different concepts.

2.6 The expanded mind eludes us

The function of expanded consciousness is largely unacknowledged and goes well beyond what we might at first imagine it to be. The expanded mind plays a major part in our ability to recognise faces; however this ability is part only of the far broader question of how we discern anything, whether a toothpick or a mountain! How do we form an image and achieve sight and understanding? I return to this in Chapter 4.

The point I wish to emphasise here is the necessity of recognising the involvement of the expanded mind. Out of consciousness and unobserved, a highly complex activity of detection and discernment is taking place that achieves results in what, for consciousness, is no time at all. Take the example of how, in thought and speech, we pluck a word from the vast compendium of words that form our knowledge of the language. How do we select the very word that carries the meaning we intend to express at a specific moment? The activity of the expanded mind is non-apparent, as is much else that we accomplish, and it functions in a manner inaccessible to conceptual consciousness. When mind activity moves from the linear progression of thought to the multiple, simultaneous activity of the expanded mind, conceptual consciousness is left behind. All further activity of the expanded mind is inaccessible and therefore, with our conscious mind, we name it the *non-conscious* or the *unconscious*. This is the way the conceptual mind views multi-consciousness! Yet, it is quite obvious that an immense intelligent activity of the expanded mind accompanies our every action and is an essential part of our awareness as a living being. To recognise an object, to sense the meaning of a word, we require the expanded mind. Simply standing upright or balancing on a bicycle demands that we integrate so many varied sensory inputs that, were it left to conceptual consciousness, we would be quite overwhelmed and would never accomplish the task. Our abilities, both intellectual and sensory, draw continuously on achievements of the expanded mind.

To understand colour vision, we need to recognise the structuring, intelligent activity that takes place outside of consciousness. Clearly, conceptual consciousness is essential in our lives but we need to be vigilant and not let it function as a blinker, restraining recognition of the full expanse of the mind's awareness, conscious and unconscious.

Having noted the distinction between conscious and non-conscious mind activity, we are equipped to enter the subject of colour vision. Chapter 3 presents the singular core process of colour formation which I refer to as the unified principle of colour.

3
Resolving Contradiction Through Colour

ABSTRACT 3

- A parallel exists between perceiving colour and depth perception. The sense of depth is generated in response to spatial contradictions; colour vision is a response to luminosity contradictions. Both 3D vision and colour vision arise out of necessity at the moment of perception.

- Contradictory brightness levels in the L, M, and S cones impede the formation of a unified view.

- To resolve the impasse, the observer generates a new mode of seeing that goes beyond registering differences in brightness. The new mode resolves the clash of conflicting views into a single, coloured image.

- This is achievable because every colour we see contains a particular combination of different brightness values, established independently by the cone sets.

- The mode of seeing that integrates conflicting views is the unified principle of colour formation.

- Wavelengths of light possess no colour and there is no direct causal relation between wavelength and colour. Attributing colour to wavelengths is a misguided assumption that needs to be abandoned so as to enable a fresh look at the question of colour.

- To understand colour formation, we need to include a third domain beyond physics and biology, that of the consciousness of the subject.
- Colour vision is engendered by the creative intelligence of the would be perceiver.
- Seeing colour is an act of defiance in the face of the impossible. The presence of colour rewards temerity with success.
- By defying the presence of a contradiction, the subject unites incompatible brightness levels to produce a visible object.
- A problem raised in Chapter 1 has not been answered: the lack of a common basis for comparing the brightness values of the three cone sets. This has to be found to enable colour vision. Chapter 8 will show how this is achieved.

This chapter describes the impediment to seeing that leads to colour vision.

Due to the differing sensitivities of the retinal cells, light made up of only part of the visual spectrum generates conflicting views. The L, M, and S cones form separate black and white views that display contradictory brightness values (as detailed in Chapter 1). As a result, the perceiver is unable to form an image. This visual impediment cannot be resolved within the framework of the physical factors that generated the problem and something entirely new is required. The needed change lies elsewhere; it comes through a transformation on the level of consciousness.

3.1 Seeing 3D

To make the case much easier to follow, I begin with an example of depth perception. This analogy is helpful because 3D vision also presents an apparently irresolvable contradiction that has to be overcome by a transformation in consciousness. Each eye forms a two-dimensional view of what is being looked at, but the views do not match due to our eyes' spatial separation. The observer is faced with two distinct views that cannot be unified through superposition.

Try the following: shut one eye, hold out your arm in front of you raise a finger and note what lies directly behind that finger. Then, without moving your position, look at your finger through the other eye and note what now lies behind it. There is a lateral shift in the background and the two images are different. The contradiction between the views our two eyes produce should impede our ability to form a single intelligible view, but it does not.

A transformation in understanding opens up a whole new way of seeing. From understanding and experiencing a two-dimensional projection to grasping the dimensions of nearer and further, there is a shift in consciousness. It is this shift that enables the transformation of two views into one single view by introducing a new quality of seeing. Creating the experience of three-dimensionality enables the observer to break through the contradiction. In doing so, the confrontation disappears and the perceiver moves beyond the conceptual framework that determined monocular vision.[10]

A striking parallel exists between the transformation that generates 3D vision and the transformation that generates colour vision. Both are incompatible views that need to be unified. With depth vision, resolution comes through creating a changed sense of space; in colour vision resolution comes from a novel way of seeing contradictions in brightness.

---

[10] I realise this description reverses the usual way this process is presented. Instead of saying that the world is in 3D and that the separation between our eyes enables us to calculate the different angles of view present in what we see, the opposite is claimed. It is the observer that generates three-dimensionality and applies it to the world. Other examples will follow that show that perception can be understood, not as the reproduction of what is 'out there', but as a creative act of the observer.

3.2 The principle of colour formation

Let us consider the appearance of colour and the function of the three sets of cone cells. For this I return to the description of a hillside with flowers that L, M, and S cones register as different in brightness, but colourless.

The would-be perceiver intends to see a single coherent view of the hillside. Is there a way of seeing that can unite the differences in the brightness registered by the three cone sets? Take a moment to reflect and you will realise that colour allows an object that appears dark to the S cones, to appear bright to the L cones *at the same time.* This produces a single unified impression without contradiction. The blue gentians appear blue because this particular hue contains just the right blend of differences to unify the lighter grey of the S cones with the darker grey of the L cones. In a similar way, seeing the rhododendrons as pink unifies what is darker grey to the S cones with what is brighter grey to the L cones. The intermediary M cones lean slightly to one side or the other.

Colour has the ability to unify the contradictions produced by different shades of grey, that is, between different, concurrent levels of brightness. With a shift from a consciousness of brightness in terms of lighter and darker, to a more complex incorporation of brightness differences, the three incompatible renditions become a single coherent image. No longer seen in terms of lighter and darker, they now register in the perceiver's freshly formed language of coloured hues.

3.3 A pause to identify the unified principle of colour

What the previous section described is the thesis and essential content of this book, the fundamental principle of colour creation. It is so outrageously simple that I have paused so that it will not slip by unnoticed. It is the unified principle of colour formation.

3.4 Some conclusions already emerging

Q: *Where does colour form?*

A: In our *mind* – the word we use to signal the seat of our ability to acknowledge discernment.

Q: *When does colour form?*

A: When we are faced with a visual impasse and there is the need to create a solution.

Q: *How does colour form?*

A: Through the elaboration of a novel visual response generated by the observer.

Q: *What are colours made of?*

A. Colours have no physical components. They form when the observer produces a visual image (see the two analogies of a mirage and of depth vision in the Introduction).

Q: *Do colours exist?*

A: Yes, of course they do – just as 3D depth, beauty, and meaning exist.

Q: *Where do colours reside?*

A: In the mind of a living subject (*mind* is defined in Chapter 2).

Q: *Out of what do colours form?*

A: Out of the unlimited possibilities that are the source of all that becomes perceptible. This potential becomes manifest in the mind of the individual generating the perception.

The potential for colour is part of all that can be. The intent to see conjures this potential into actuality. We can understand all that forms our experience of the world as one particular rendition of what can be. Colour, like depth vision, is an enrichment of our personal experience and adds new ways for beauty to manifest in the world. Transforming what was colourless into colour is one of many creative acts we perform in order to sustain our view of reality and to generate a coherent, stable view of the world and of our place in it.

These remarks complete this chapter, but I will add several points to clarify what has been said so far.

3.5 Light is colourless

I return to a point that may still be troublesome to the reader: the fact that colours do not correspond to different wavelengths of light and, in fact, they are absent in all wavelengths. If this continues to shock, it is most likely because we find it difficult to shake off an entrenched assumption that links colour to wavelength. We need not think of colour in objective terms as something present in physical reality; instead, we can recognise colour as generated within the act of perception. Colour arises out of necessity at the moment of perception. To see colour attests to the achievement of a breakthrough in understanding. It reflects intelligence on the part of a being that has the power to move beyond certain limitations in visual perception and generate a new and different mode of seeing. Such an achievement is not contained within the physics of light. All colour phenomena touched upon in these chapters can be accounted for without attributing colour to wavelength.

In the supplementary chapters of Part II, the Benham top is a stark example of transforming a black and white pattern into a many-coloured image. It shows how colour formation can be described solely from the conflict between the views of the cone sets, without any recourse to wavelengths. But we have a way to go yet before reaching that example.

3.6 Black and white images

Having spoken only of colour formation, we should, perhaps, turn the question around and ask: in situations where we see a black and white scene, why does it remain in tones of grey and not transform into colour? The reason is simply the absence of conflict. The L, M, and S cone sets concur in the brightness ratios they register because full-spectrum light does not create conflicting views between them. In full-spectrum light, all three cone sets see what is bright as bright, and when it is dim, all three see a dimmer image and no conflict arises in the brightness levels recorded. There is thus no call for a resolution through the transmutation of the brightness values into colour.

3.7 The unseen content

One question is left: where are the three black and white views of the scene that are produced by the cone sets, since we do not actually see them? The answer is that there is more to the mind than what we consciously observe. As presented in Chapter 2, the mind functions on a non-conscious level – an obvious fact that we repeatedly seem to overlook! We have conscious thoughts that we know we are thinking and beyond this a lot is happening on a mental plane that does not enter consciousness. Examples abound that show the mind treats a wealth of sensory information out of consciousness.[11] For instance, in the visual process, the saccadic movement of the eye is essential for gaining sight of an object, but this activity is not processed in consciousness and we remain oblivious of it. Other examples will follow.

---

[11] The whole topic of the conscious and the multi-conscious mind is not included here. I have written about it in *Making Sense*, Chapter 1, and elsewhere (see footnote 3, p. 4). I only include here what relates to the process of colour formation

3.8 The unsolved issue

But wait! I have moved too fast in this explanation and have left out a crucial factor. Everything thus far is merely a preliminary description to permit the reader to enter into the logical framework. In actual fact, there is, as yet, no overall description of colour formation because, as you recall (in the analogy of the orchestra), the range of brightness of one set of cones cannot be compared with those of another set. The L, M and S cones (Chapter 1) do not share a common scale that would allow the brightness levels in one cone set to be compared with those in another. This incongruence required to keep them from blending into a single grey representation. But, for sight to be achieved, this dissociation has to be overcome. How, finally, do the three separate views interrelate?

A response to this comes in Chapter 8. Before then, there are other items to consider. The next chapter presents the momentous process of image formation. Understanding the process of colour formation requires an understanding of how we form a visual image.

4

Image Formation

ABSTRACT 4

- Image formation, although frequently overlooked in the context of seeing colour, is an essential part of the process leading to colour vision. To see something requires the elaboration of an image from diverse sensory impulses.

- This is achieved by recognising relationships: forms, positions, luminosities and so on.

- Only when ratios are established across all that is looked at does visibility become a fact.

- Relationships are mutually determined by what is present in the field of view and no outside reference for distance, angle, brightness and so on can be employed as no such reference measure is available.

- The realisation of an interrelationship can be thought of as a wave of integration passing through all that forms what is seen. In fact, seeing may well be the experience of such a wave.

- Through this activity, the expanded mind enacts what is needed to produce visibility by generating a unified view that can become conscious.

- The achievement of an image is necessarily a momentary phenomenon, for it has to be released to allow a subsequent image to enter consciousness.

- From the fact that we can perceive a rapid sequence of changes, such as a moving object, we can conclude that images we see are refreshed and updated in milliseconds.
- The formation of an image is accompanied by the implicit understanding of 'not me', by the recognition of every object as 'out there', that is, as not being me.
- Each cone set produces a separate image of what is looked at.
- In the case of colour vision, the conflict in brightness between the three independent views is preconscious. Once unity is achieved, the result enters consciousness as a coloured image.

How does a visual image form? We know that stimuli reach the eye and are relayed through the organism's visual system. But what is required for the stimuli to become an image? An answer is essential if we are to understand sight and colour vision.

4.1 The uncanny moment that enables vision

The formation of a visible image is at the crux of the process that yields colour. The image we see is the object we perceive. It is not at all the same as the reception of impulses through the optic nerve. They have to be interlinked in a coordinated fashion for something to become visible. An act of integration must take place in which numerous factors blend to create a moment of perception, thus generating consciousness of the object we see. This moment in the process of perception is the pivot I wish to emphasise. The production of an image holds the answer to questions about the role of the subject, the establishment of consciousness, and the presence of an object exterior to the subject. All of these are involved in generating colour vision and the hues we see.

4.2 Forming a visual object

Because we assume that what we see is 'out there' in the world and because we know that the light reflected from objects we see enters our eyes, we easily overlook the fact that the visual image is not formed directly by what enters the eye. Our sensory and intellectual abilities have to accomplish this by integrating the stimuli that impinge upon our visual system. But where and when are the stimuli transformed into the image we see, and what motivates this process?

There is a relatively simple answer that does not require a laboratory, with an experimenter and test subjects. When we take the time to think carefully through the process, the evident conclusion is that a visual image, just like any meaning that forms in our mind, results from an act of integration that is the precursor, as well as the initiator, of consciousness. I will briefly describe what this entails.

Every object and every scene we see is made up of parts. To take a minimalist example: to see Venus shining brilliantly in the predawn sky is to recognise a point of light on an extended, less illuminated background. A relationship is involved. Everything we see is made up of a composite of relationships. A scene we look at contains different forms and in recognizing their relationships we recognize what we see. To see a square as a square, for instance, the perceiver takes note of the relative length of the sides of the figure, of the angles they produce and of the fact it is a closed figure. Without taking account of these, the perceiver would not see the object as being square. This information is taken in by the mind non-consciously, and possibly also consciously. To see a fuzzy patch or a squiggle or anything requires the identification and acknowledgment of relationships. This invites the argument that what we perceive are not objects but the simultaneous, momentary presence of relationships. We see ratios! Sight is an intuitive act. Leaving the implications of this to the side, the inevitable conclusion is that achieving integration is central and, in order to see a full scene, innumerable relationships must be deciphered and brought into a unified view.

It is well known that our visual system does not simply 'photocopy' what we set our eyes upon. The image is created in our mind. Without moving into the more fundamental question about what is real, we can recognise that it is we, the subjective viewer, that generates and forms the images that enter our consciousness. The outlines of objects are brought into relationships that permit us to recognise what we see, and much more than outlines are integrated to form the image. The act of seeing seems so simple that we imagine we do this consciously in the blink of an eye, but actually the task is immense. As discussed earlier, in order to form the view of a scene, we have to establish an overall ratio of brightness levels of everything in that field of view. There is more to it than this. In seeing an object, a flower, a person, a mountain stream and Venus in the predawn sky, we experience each of these as a full understanding, which goes beyond the dictionary definition of that object. It is both objective and personal. The object's meaning is sourced out of our personal learning, emotions, likes and dislikes; it is imbued with memories and experiences from a lifetime and beyond.[12] There are other implications, such as innumerable, personally relevant connections and meanings including symbolic overtones that generate our sense of the object. All this comes together at the moment of recognition. A simple concept – island, fragrance, strong – carries specific personal meanings. The wealth of information present at the instant we recognise a word is enormous, and so too is the wealth of information present the instant we acknowledge what we see.

When we consider how all of these meanings congregate and unify in the object we see, we begin to appreciate what is actually happening at the moment we recognise a visual object. The whole enterprise

---

[12] Instinct is one example of overreaching immediate lifetime experiences.

36

is elaborated without our knowing it. Physiological, psychological, emotional, even extrasensory factors integrate in what we perceive, without any conscious search on our part and without us expressly gathering the innumerable attributes and inserting them in the object we perceive. We are conscious of the result and not of the process.

The result is always a single view, a single object, a unified some-thing. Yes, a something that is not me. It sustains the experience I have of its *being other:* what I see is other than me, and I am other than what I see. Whatever the diverse impressions that impinge on our sensory system, the result is the perception of a whole that constitutes an independent presence. An object manifests, whether as an imprecise impression or as a clearly recognisable thing. Much like a child blowing bubbles, we release entities, bubbles of whatness, each an integrated whole.

4.3 From physical process to image

We can add a further dimension. We know of the electric impulses and chemical transformations that stimulate our nervous system, but we do not experience them as flashes and pulses but as the image of a flower or a friend. This transformation involves the same, unseen process of integration. To produce a visual image and see, requires a massive integration of physiological factors, which is also accomplished subliminally. We observe only a result that bears witness to what must have happened beforehand, beyond the reach of consciousness. All this leads to the conclusion that any object or scene we see is a momentary crystallisation of diverse relationships.

In terms of what was said in Chapter 2, this process takes place in the realm of the mind that is not conceptually conscious, that is, in the expanded mind not limited to linear processing. The expanded mind is the seat of vast imperceptible activity that culminates with integration. Once a visual object is formed, once a single view of an object or a scene is realised, it can be delivered to the conscious mind. It is then that we, the subject, become conscious of the visual object.

4.4 The duration of a visible image

With this comes the question: when the view crystallises and we recognise what we are looking at, how long does that image persist? Clearly, if the image lasts longer than a moment, the perceiver would be left behind, stuck with an image of what was, unable to follow what is ongoing. The image that forms must be released in order to see change and movement, to perceive the flight of a bird or a leaf trembling in the wind. The duration of an image must be exceedingly brief. The view needs to be recreated every tiny fraction of a second.

From all this we can only conclude that the world we see is formed through an act of integration that takes place out of consciousness. The expanded mind brings together diverse attributes that interrelate to produce a perception. The resulting image enters consciousness and is continuously refreshed.

On a more philosophical plane, our perception of reality is being recreated moment by moment. It would not be right to conclude from this that the world is made up of fleeting moments with imperceptible gaps between them, nor to conclude that time is a non-continuous series of separate moments. What it does suggest is that the concepts through which we understand our world are formed in this manner. The act of perceiving is discontinuous and results from interrupted momentary perceptions.

4.5 The moment of integration

The point of this chapter is to emphasise one basic fact: integration is the precursor to every act of seeing. Integration is an essential step in creating an image, whether black and white or coloured. For colour formation, the conflict and resolution that happens on the level of brightness values is essential. The drama takes place out of consciousness in the expanded mind, and the resolution of the conflict is the recognition of a coloured visual image. To undertand the process of colour formation, we must necessarily include an account of the formation of visual images.

4.6 Conclusions

1. To see something requires the production of an image from multiple sensory impulses.

2. The transformation of sensory input into a visual image is achieved through the recognition of relationships: forms, positions, luminosities. Relationships are established and only then does what is looked-at become visible.

3. These inherent relationships form implicit, internal, value scales. There is no unit measure, such as a centimetre or a lumen, as their function is replaced by simultaneous mutual relations between all that is present in the visual field. (See also Chapter 8)

4. Each cone set produces its image of what is looked at. These three images are brought together. If their brightness values concur, the image we perceive is in the range between black and white. If to generate colour vision, they must be in conflict and yet have an underlying structure. The sensitivity range specific to each cone set is the structure that underlies the three different views, each responding to all that is present in a single field of view. Then, on the subliminal level of mind not available to conceptual consciousness, the three separate views encounter one another, interrelate, and produce a single coherent visual content.

5. Formation of a visible object results from an act of mind that unravels the discrepancies into a single momentary view. To ignore

this fact and to note only the cause-effect details of neurological processes would be a strange omission. It would be like explaining why living creatures eat by detailing the mechanism of the digestive process. Without the desire to eat, the process would not be set in motion. There would be no digestive process without the sense of purpose present in hunger. This expression of purpose manifests on a separate plane, distinct from the stages of the process of digestion. These thoughts, central to a full view of life processes, lead away from the immediate topic into other areas of inquiry. I therefore return to the previous statement and conclude with the fact that the formation of an image and the recognition of a visual object is achieved through a process of integration.

4.7 Looking ahead

It is now time to consider colours themselves. Chapter 5 is a brief study of subtractive colours and their relationships.

Figure 5. A rainbow (Photo: Siw Backas, Helsinki, Finland)

5

Colours and Their Relationships

ABSTRACT 5

- The familiar rainbow spectrum of colours is formed by the juxtaposition of two separate projections. I refer to them as *half-spectra*. One is the violet to light blue range, and the other is the dark red to yellow range.

- Where the ends of the half-spectra encounter one another, we see green on one side and magenta on the other.

- The two distinct half-spectra are a fundamental fact of colour vision. Recognising this simplifies the description of certain observations as colours can be recognised as belonging to either the red half-spectrum or the blue half-spectrum.

- The colour wheel, known for centuries to artists and scientists, is formed when the two half-spectra are joined in a circle.

- In approaching green, both sides of the colour wheel (yellow and light blue) become brighter, yet green itself is not part of this brightening process. So too, in approaching magenta, the colours of the circle become darker (dark red and violet), but magenta itself is not part of the darkening process.

- On the brighter side of the colour wheel, the blue half-spectrum and the red half-spectrum approach a common wavelength from opposite directions. The longer wavelengths shorten; the shorter wavelengths lengthen, merging in the mid-wavelengths of the visual spectrum.

- Where they meet, we find green, a colour that emanates a calm quality.
- On the opposite side of the colour circle, the wavelengths of the two ends stretch apart in contrary directions. On the blue side, the shorter wavelengths shorten even more; on the red side the longer wavelengths lengthen.
- On the colour wheel these two ends meet to form magenta, a colour with a wild, energetic quality.
- The complementary colours, opposite one another on the colour wheel, are formed out of opposite brightness levels. For this reason, when brought together, they annul one another and lose their colour.
- Colours that are not exactly opposite on the colour wheel are not fully complementary, so when they are mixed, they do not counterbalance exactly and a murky colour appears that is in the direction of colourless.
- Analogue colours (colours that are adjacent on the colour wheel) show a gradual brightening or darkening of the neighbouring colour; not so, however, with green and magenta.
- Even though there is a gradual progression from adjacent hues into and out of green, and into and out of magenta, these two colours are distinct and, unlike other analogue colours, are not predictable from the neighbouring hues.
- The relationship between the colours of two half-spectra are apparent from their place on the colour circle and we recognise these relationships visually. However, when described in terms of wavelength, some relationships are far from obvious, such as the appearance of green, a distinct new colour, in the midrange of the visible spectrum.

5.1 The colours of the spectrum

When light passes through a prism, two separate colour displays are produced. We often see them side by side and thus are accustomed to considering them as a continuous rainbow of colours. In actual fact, the familiar rainbow is made up of two separate projections, which I will refer to as *half-spectra*.

We habitually consider these two half-spectra together as being *the* colour spectrum. When a beam of light passes through a prism, one half-spectrum is produced alongside one edge of the beam, the other half-spectrum is produced alongside the opposite edge. We only see this when the beam is wide enough for the two edges to remain at a distance from one another, leaving a white central part in between.[13]

A simple way to observe this effect is to stand before a bright window and hold a prism horizontally in front of your eyes. When the

Figure 6. Black and white stripes (on the left) when seen through a prism show the two half-spectra separately (on the right). When the space between the stripes is reduced, the half-spectra meet and green appears.

---

[13] The white central part is not questioned at this point, I simply wish to describe what is seen.

upper edge of the window frame is refracted through the prism, we see the colours of one half-spectrum, for example, dark red above ending in light yellow. Turn the prism slightly to look at the lower edge of the window where the dark frame below encounters the light above and you'll see the sequence violet to light blue – the other half-spectrum. This shows that the two processes are clearly distinct.

If the window is narrow and its top and bottom edges are seen in the same view, the two half-spectra may touch and where they meet, you'll see green. In this case, the two half-spectra are joined in a continuous sequence and you'll see a full spectrum of colours – or almost, that is, for magenta has been left out.[14]

In the case of a rainbow, the edges of the raindrops are close together so when light is refracted as it passes through them, the two half-spectra touch one another and form a continuous band of colours. The red to yellow sequence moves into the blue to violet sequence and where yellow and blue meet, we see green.

The two distinct half-spectra are a fundamental fact of colour vision.[15] Without this underlying structure, there would be no colour circle, no complementary, primary, or secondary colours. A full discussion of this process and its significance leads well beyond where I

---

[14] Magenta is formed when the other two sides of the half-spectra touch one another, not yellow and light blue as in this example, but dark red and violet. When we observe two separate prisms that bring these sides together, the magenta that appears is stunning – I gasped the first time I saw it. It is a powerful effect that no colour pigment is capable of reproducing.

[15] This fact is not often given prominence, probably because of the association of colours with wavelengths. Light is understood as a segment of a continuous electromagnetic spectrum formed by different wavelengths set in a linear sequence from 380 nm to 780 nm. Colours are attributed to different wavelengths, for instance, green is attributed to the midsection of the spectrum (495-570 nm) and not to the encounter between two separate progressions. In such a viewpoint, there is no place for separate complementary processes of colour formation, at least not without supposing concealed neurological processes in the brain. I feel confident that a reader with this linear interpretation of colour will, before turning many pages, reach a very different realisation.

wish to go at present. In this context, I merely point out the role of the perceiving subject in colour production and answer some basic questions about colour formation in relation to the two half-spectra. However, in Chapter 10, there is a suggestion of what lies beyond this first level of understanding.

The distinction between the two half-spectra simplifies the description of diverse colour effects, and often the significant fact is to recognise whether a colour falls within the blue-violet or the yellow-red range. This will be done on occasion in the chapters that follow.

Recognising colour relationships as a circle of gradient hues comes from our direct experience of colours. On the basis of wavelengths alone, these relationships are baffling. A bridge between what we observe and the wavelength description emerges when the reason for the colour wheel and its complementary halves is investigated.

This is exactly what is needed to enable a rewarding dialogue between artist and scientist.

5.2 Five points already covered

1. The three cone sets of the retina differ in their sensitivity to light of different wavelengths.
2. When the variation (ratios of brightness) between what appears light and what appears dark is the same for the three cone sets, we see a black and white image; this happens when the light reaching the eye is full-spectrum sunlight.
3. When the light is not full-spectrum, the differing sensitivities of the L, M, and S cones result in contradictory views. The brightness relationships established in one cone set will not be the same as those established in the other two cone sets and the views formed by the different cone sets will not cohere.
4. When contradictory views are seen as colour instead of merely as brightness (i.e., black and white), the contradictions are resolved

and the differences in brightness are brought together in a single unified view (as described in detail in Chapter 3 and elsewhere).

5. Each colour hue embodies the resolution of a particular brightness conflict. The progression along the colour wheel from one colour to an adjacent hue is the transition needed to accommodate incremental changes in the brightness ratios of the three cone sets. A minor change in luminosity ratios alters the brightness relationship between the three cone sets and a corresponding adjustment in hue accommodates the new values into a single view.

Bear these points in mind as we look at the colour relationships which involve the subtractive mixing of colours. Only these relationships are discussed in this book; the mixing of coloured lights, known as additive mixing, is set aside for a later study.

5.3 The artist's colour wheel

Figure 7. The colour wheel is made of the two half-spectra curved into a circle. At the two ends where they join, we see green and magenta.

The sequence of colours in the red half-spectrum and those in the blue half-spectrum can be placed end to end to form a circle known as the colour wheel (see figure 7). This produces a graded sequence of colour that makes intuitive sense. The circle makes it is easy to recognise relationships between colours. For example, where the two colour sequences, or half-spectra, meet, new colours form − green on one side and magenta on the opposite side. Adjacent colours flow together to form a gradually changing sequence, for instance, from red to orange to yellow. Placing colours in the form of a wheel also sets complementary colours diametrically opposite one another. The complementarity is expressed by the fact that opposite colours, when mixed together, annul one another and the result is black.

When colours are roughly of equal intensity on the colour wheel, one side of the circle is brighter[16] and the opposite side darker. The colours of the two half-spectra become darker towards violet and towards dark red; they are darker in the direction of magenta, but magenta itself is not part of this sequence. On the opposite side of the circle, the colours become brighter towards light blue and yellow; they are brighter in the direction of green, but green itself is not part of the progression. Although magenta occupies the centre of the dark half of the colour wheel, and green occupies the centre of the light half, these two colours appear independent of the dark light progression.

The above describes what we see directly; however the same phenomena can be described in terms of wavelengths.[17] On the colour wheel, the colours become brighter towards the midrange of the visual spectrum, around 550 nm; they become darker towards the two

---

[16] I employ the word *bright* without taking account of the technical distinction between *lightness* and *brightness.* Here, both these words signify what we experience as the opposite of darkness.

[17] Speaking of wavelength is a change in register. It no longer describes colours in terms of what we recognise visually as colour phenomena. Recognising the relationship between the two descriptions is a repeated theme.

extremes of the visual spectrum, where the wavelengths approach the limit of what our eyes can see. In other words, the wavelengths approach the boundaries of what we know as light.

The two ends move into infrared on one side and ultraviolet on the other.[18]

5.4 Green and magenta

The transition towards the brightest range of colours involves two contrary progressions. On the red side of the colour wheel (where the light registers more strongly on the L cones), wavelengths get shorter. The progression from red to yellow corresponds to a shortening of wavelengths.[19] On the blue side of the colour wheel (where the S cones are more stimulated), the wavelengths get longer. From violet to light blue corresponds to a lengthening of wavelengths. The two progressions merge from opposite directions towards common, shared wavelengths.

It seems significant, or should I say appropriate, that coming together to meet at the same wavelength produces an experience of calm – the quality of the colour green. This is in contrast to the opposite side of the colour wheel, where there is a movement of separation. On the blue side of the colour wheel, the progression towards violet corresponds with wavelengths becoming shorter, while

---

[18] When we consider the light-sensitivity curves of the cone sets, a reason for the gradual change in brightness suggests itself. The plot of the sensitivity range of each cone set is roughly a bell curve, low at the sides and rising to maximum sensitivity in the midsection. The greatest combined sensitivity of the curves is roughly towards the midsection of the visible spectrum. This may be part of the reason for the brightening, but there is something more. Because the overlap between the curves is greatest at mid-wavelengths, one can assume that a lower proportion of the overall illumination is in conflict at that part of the spectrum. Where there is less contradiction, the 'white light' brightens and dilutes what is seen. I leave this thought here, without following the lead further.

[19] A reminder: it is not the wavelength that is red but the lack of the other wavelengths that creates the Imbalance that the subject resolves by seeing red.

on the red side of the colour circle, the progression towards deep red corresponds with wavelengths becoming longer. The two progressions are moving apart; they fade out of visibility at their greatest separation. The colour that comes to resolve their differences is the violent shade of pink we know as magenta. Magenta seems to scream out, while green expresses harmony. I believe there are meanings to discover in the way we perceive physical processes.

5.5 Change in brightness versus change in wavelength

Two different descriptions of the same effect are presented, one in terms of a progression towards or away from the light, and the other in terms of a progression in wavelengths. I present this double viewpoint to speak to the artist within the analytical thinker, and to the analytical thinker within the artist.

You might still ask, why doesn't a description of wavelengths suffice? The answer is simple. The physics of light, conceived as a linear progression of wavelengths, does not accommodate certain observable facts. The transitions to green and to magenta take place where the ends of the half-spectra meet; however, from the viewpoint of wavelengths, there are no ends that meet! Green is associated with the middle of the visible wavelengths and no particular change is noticeable there, while magenta has no place on the gradient of wavelengths and appears when the shortest and the longest visible wavelengths combine. This is curious.[20] I understand this to imply that our visual ability is not entirely

---

[20] These effects seem illogical and probably made it difficult for physicists to deal with the colour wheel and to have a fruitful dialogue with artists. The incompatibility of the two approaches used to be more pronounced than it is today as, for example, in the conflict between Goethe and Newton. The conflict has become less stark, not because of progress in understanding but because, to a wide extent, art texts have submitted to dominant scientific terminologies and explanations.

determined by the light that reaches our eyes. The achievement of perception requires a completion through the intervention of a living being, of the individual being that enacts the processes of integration that supplement physical and neuronal interactions.

5.6 Two distinct colour progressions

Now for some further details on the brighter and darker sides of the colour wheel and the wavelengths involved.

Green appears when the longer wavelengths of the blue half-spectrum (to which the S cones are more sensitive) encounter the shorter wavelengths of the red half-spectrum (to which the L cones are more sensitive). We see green where they come together in the mid-wavelengths of visible light, that is, in the centre of the brighter side of the colour wheel. The progression of colours on each side of the colour wheel merge and point in the direction of a meeting place. And what is that direction? Both sides progress towards that which is brighter. They move towards the light.

What about the other meeting point, where magenta appears? That movement is of separating wavelengths, where the ends of the half-spectra move out of visibility towards opposite extremes of the spectrum. The progression we observe is towards the dark side of the colour wheel, into violet and darkening red. Each half-spectrum moves away from the light and towards darkness, only to yield a final flare of magenta before stepping into the void where light is no longer visible.

Magenta (although possibly brighter than the surrounding colours) holds the centre of the side of the colour wheel that moves in the direction of darkness; green (although darker or more present than the surrounding colours) holds the centre of the side of the colour wheel that moves in the direction of light.

5.7 Complementary colours

Colours that are opposite one another on the colour wheel are complementary. Mixed together, they annul the presence of colour. But there is more to this description.

For example, the complementary colours orange and blue are positioned opposite one another on the colour circle. We know that orange appears in order to resolve an imbalance produced when longer wavelengths stimulate the L cones more than the S cones. Its blue complement is produced by the observer to resolve the contradictory views of S cones brighter than L cones. When we mix orange and blue together, what happens? We combine the two separate ratios of brightness. The excess of light from the L cones (in the case of orange) is combined with the excess of light from the S cones (in the case of blue). The paucity of light from the S cones (in the case of orange) is combined with the paucity of light from the L cones (in the case of blue). Together they create a balance in which all the cone sets register a similar level of brightness, which means that the luminosity differences between the cones are equalised, thus rectifying what was contradictory. As in the case of full-spectrum light, no remaining conflict troubles the achievement of one unified view. Colour is not called for and vision returns to a grey scale between black and white.

It is important to note that complementarity has to do with the sensitivity curves of the cones. It is not a matter of measuring the brightness values on an absolute scale but as relative to a norm.

Our eyes are adapted to the solar spectrum and to the particular ratio of luminosities that, for our vision, is recognised as being in balance.

So, what happens when colours that are not exactly opposite (or complementary) on the colour wheel combine? If deep red is mixed with dark blue, for example, they do not fully annul one another. The brightness of one cone set is not exactly matched with a colour that is not positioned exactly opposite it. The result is a murky colour. The wavelengths do not fully annul their differences and the resulting hue is less coloured. It lies in the direction of colourless. This describes what happens when colours from different half-spectra are mixed.

Of course, green and magenta, also of different half-spectra, have a different story that will be elaborated in what follows (5.9). In truth, the descriptions in this chapter are simplified and a deeper level of understanding awaits you in Chapter 10.

5.8 Analogue colours on the same half-spectra

The description above has been about mixing colours that lie in different half-spectra, that is, on opposite sides of the colour wheel. Now we come to the question of mixing colours on the same side of the colour wheel. The result, in this case, is not the same deadening of colour. Light blue mixed with violet (both of which are on the blue side of the colour wheel) or deep red mixed with yellow (both on the red side of the wheel) do not produce a hue that goes in the direction of colourless.

There is a gradual change of colour around the colour wheel that shows a sliding scale of hues. Any one hue appears as an intermediary to its bordering colours in two possible progressions, either towards brighter and warmer colours, or towards darker and cooler colours.

The colours on each half-spectrum are achieved by adding a percentage of *non-discordant* light to the red or the blue. Non-discordant refers to light that does not produce a contradiction between the cones and thus appears white. When mixing paints, adding white to red or to blue can bring the increase in brightness.

However, on the colour wheel, more than a change in brightness is involved. Were it only brightness, we could expect the red half-spectrum to change to pink, as red does when we add white.

The colours we see are the result of the ratios of brightness of three cone sets. To accommodate not two but three variables requires the creation of a further colour change and this is the function of yellow. It appears on the red half of the colour circle. Why is there not a similar new independent colour that enters on the blue side of the colour wheel? More contrast in brightness is recorded on the red half-spectrum because two cone sets, L and M, differentiate brightness values. On the blue side, only one cone set, the S, is predominantly sensitive to that half-spectrum. There is also the fact that there are fewer S cones compared with the number of L and M cones. The presence of brightness differences figures more strongly on the red side of the colour wheel than on the blue side that, consequently, does not call for a similar colour change.

5.9 Analogue colours on opposing half-spectra

I have described the graded progression within each half-spectrum. But how should we describe what happens when the ends of the blue and red half-spectra meet on the colour wheel? We find a shift to unexpected colours, to green and luminous magenta. Why unexpected? Because these two new colours cannot be approached by adding either white or black to the adjacent colours. Were we not so accustomed to mixing blue and yellow, we would not be able to predict that green will result. This is clear when we take a parallel example from the additive mixing of colours. In a comparable situation that is unfamiliar to most people, it is quite a surprise to discover that the combination of red and green produces yellow.

Returning to subtractive colours, the transition from yellow to green is not a gradual darkening or brightening of the yellow and neither is green a darker or brighter shade of blue. Even though we observe greens that tend more towards blue or more towards yellow, the colour green is distinctly 'other' and different from the colours that mix and form it. This suggests that the transition between the two ends of the half-spectra (where we find green and magenta) is a process other than that which functions with analogue colours within the colour wheel. It shows that, in terms of what we see, we need to recognise an event that does not feature in descriptions that associate colour with wavelength.

5.10 Describing not explaining

The emphasis here is on the involvement of the individual perceiver and how this points to the fact that the experience of colour is not a secondary side-effect of physical events but is the source of what becomes visible. Colour enables sight. This is its immediate function. In this, the unified colour principle diverges from what is commonly presented in contemporary colour theory.[21] The approach I follow requires that we step aside from established interpretations and formulate afresh the questions which guide our observations. My

---

[21] For instance, there is a claim that colours, in unseen ways, oppose or enhance one another. This 'opponent process' was proposed in 1879 by Ewald Hering to answer unexplained problems in colour relationships and it did so by attributing excitatory and inhibitory responses to certain colours. It is a hypothesis based on unproven assumptions about neuronal activity that determines colour interactions. Still today, this approach is given as an explanation of colour relationships. The unified colour principle, in contrast to the opponent process, does not encounter these colour relationships as problematic, and thus does not require a new hypothesis about colours interacting with each other. In fact, the logic behind such interactions is questionable when one realises that colours have no independent presence and form momentarily in the conscious mind of the perceiver. Even so, much of colour research has been set to the parameters of the opponent process although, basically, it is only a hypothesis. C. L. Hardin pointed out that the opponent process is an idea with no causal basis. He wrote, 'The incompatibility (between the members of an opponent pair) is the basic feature of the model, but the model does not, of course, explain it.' C. L. Hardin, *Colour for Philosophers, Unweaving the Rainbow* (Indianapolis/Cambridge: Hackett Publishing Company, 1988), 121; *Ewald Hering, Zur Lehre von Lichtsinne* (Wien: Gerold Verlag, 1878).

aim is to describe the process that enables colour vision, not to explain what colour is. Why the potential for colour exists and what is the significance of colour are questions that lead to mysteries well beyond the scope of this of this work. Presented here is an exploration of the process that enables colour vision and its implications for an understanding of perception.

5.11 Looking ahead

Before going further into a description of colour processes, I shall consider two experiences of colour that illustrate the application of the unified colour principle. Chapter 6 presents the curious case of coloured shadows, and Chapter 7 considers the case of coloured after-images.

6
Coloured Shadows

ABSTRACT 6

- The phenomenon of coloured shadows is a particular case of the principle of colour formation set out in Chapter 3.

- This typical example of coloured shadows concerns a scene illuminated by a coloured lamp, in which an object casts a shadow. With the addition of a second white lamp, the original shadow changes from grey to the complementary colour of the coloured lamp.

- A closer look at what is taking place describes the view formed by the different cone sets and the contradiction in brightness relationships produced in the area of the shadow when the second light is added.

- A resolution of the conflict in the shadow is achieved when the observer transitions to colour vision and sees the shadow as the specific colour that unifies the observed contradictions.

This chapter steps back from fundamental considerations of visual perception to describe a surprising manifestation of colour, the case of coloured shadows. It is an example of the principle of colour formation as presented in previous chapters. I open with a section on prismatic colours.

6.1 Coloured shadows

The phenomenon of coloured shadows is well known. When an object is lit with a single lamp of a certain colour, for instance orange, its shadow is dark and colourless. When a second lamp is positioned to the side of the first, that same shadow receives some illumination.[22] This addition brings a surprising change as the shadow is then no longer dark grey or black, but appears blue, the complementary colour to orange.

Figure 8. A scene is lit with a yellow light and an object casts a shadow (to the left). When a second fainter white light is switched on, as in this photo, that same shadow (to the left) turns blue.

But this is not the full story. If we only look at the shadow, without seeing what surrounds it (which we can do by looking through a narrow

---

[22] To show the principle involved, I discuss only the shadow cast by the initial orange light.

tube aimed just at the shadow), there is no change and the shadow continues to look dark grey! It seems strange that the same shadow appears both black (through the tube) and coloured (when the field of view is not reduced). Which view shows the real colour of the shadow, the dark grey or the blue?

In the context of what has been presented so far concerning colour formation, namely, that colour arises out of the immediate situation before the observer, there is no ontological distinction to be made between the two views. Both are equally present. We see genuine colour in both cases. The colour that we see is the true colour, and yes, the same shadow is grey when seen one way and blue when seen another way. There is no contradiction because in each instance, the colour is what it appears to be. I can well imagine this explanation inviting the response, 'But surely the same shadow cannot be both blue and not blue, both grey and not grey?'

This question brings the answer forward: the colour does not belong to the shadow; it is generated by the observer. The disparity between the two views of the shadow is not the puzzle, rather, what needs elucidating is why the shadow appearing grey in one situation resolves a conflict that is not present in the second situation, where that shadow appears blue and vice versa. I will trace the process of how this happens.

6.2 Colour formation in the shadow

The difference between seeing the shadow on its own or within its surroundings shows that the change between grey and colour clearly depends on the presence or absence of the area surrounding the shadow. The appearance of colour is not determined by the brightness of the shadow alone, but of the shadow in relation to its surrounding area. This implies that to understand what is involved, it is necessary to consider the ratios of brightness across the field of what is seen and not merely the brightness of the areas themselves.

To restate the situation. A scene is lit by an orange light and an object obstructing the light casts a colourless shadow. With the addition of a dim white light, the scene shows hardly any change, but the same shadow (cast by the object blocking the orange light) now appears blue. I will consider this situation from the point of view of the different cone sets.

First, the effect on the retinal L cones. The addition of the second dim white lamp has little effect on the ratios of brightness of the L cones or on the overall scene. There is a slight increase in brightness due to the longer-wavelength component of the added white light. The shadow is illuminated slightly and the rest of the scene has a slight increase in brightness. The ratio of brightness between the brighter surrounding and the darker shadow goes through no significant change.

However, the addition of the second lamp introduces a significant change in the ratio of brightness values in the S cones. Before the addition of the second lamp, the scene is lit with longer-wavelength light to which S cones are insensitive so no ratios of brightness have formed. Both the shadow and the rest of the scene are dark and are not called to respond to any incoming light. The addition of the second lamp activates the S cones and they register the shorter-wavelength components of the white light. The S cones now record a faint brightness for both the now illuminated shadow and for the surrounding area, which are viewed as being of equal brightness.

From this, we note that the presence of the second lamp introduces a contradictory view of the shadow. To the L cones the shadow appears darker than its surroundings. To the S cones it appears equal in brightness to its surroundings. This discrepancy can be overcome, as has been shown, when the scene is no longer viewed as dark and light (black and white) but incorporates the shift in consciousness that allows for colour vision.[23] What is considered darker to the L cones unites with what appears lighter to the S cones to form a single, unified, coloured view of a hue in the blue half-spectrum.

As to why the overall scene that surrounds the shadow appears orange, I'm sure after the previous chapters this is clear. In the scene illuminated by the orange light, the L cones register it as bright in comparison with the shadow. The S cones receive little or no input from these areas and are not activated by light. With the L cones registering light and the S cones not registering light, the contradiction is resolved into unity when seen as orange. With the addition of the second lamp, the underlying ratio between light and dark in the surrounding area does not change.

Finally, we consider the coloured shadow alone, seen through a tube. It should be self-evident by now why it appears dark grey. I'll restate it anyway. The shadow seen alone without the surrounding area, receives only the full-spectrum light from the second faint lamp. The L, M, and S cones respond equally to the brightness of that light. No conflict arises and the shadow does not change its colourless appearance.

6.3 The same process active in different phenomena

This description of coloured shadows is applicable to all situations. For example, if the same scene is lit not with an orange light but with a blue light, the object again casts a dark shadow. When an additional white light is added, the change in the ratio between the scene and the shadow does not change for the S cones; however, for the L cones there is a significant change. The L cones now register

---

[23] What this entails is described further in subsequent chapters.

equal illumination between the shadow and the general scene (exactly parallel to the situation described above for the S cones). The result is that the shadow appears darker than its surroundings to the S cones but to the L cones the shadow is equally as bright as the rest. Thus, in the presence of the second lamp, the brightness relation of the shadow relative to its surroundings is: the L cones register the shadow as brighter, while the S cones register it as dimmer or dark. Resolution comes when the observer transitions to colour vision and the area unifies into a hue within the red-yellow half-spectrum.

Each half-spectrum, red or blue, offers a range of possible hues, the hue that appears depends on the exact brightness values involved. On the red half-spectrum, a saturated red colour appears when there is a stark difference between the ratios observed, while orange or yellow appear when the contrast of ratios is less pronounced. In a parallel fashion, when the ratios producing the blue half-spectrum differ greatly, we see a colour towards violet, while with less difference in the ratios, we see blue or cyan.

A full description would also include the role of the M cones. I have not mentioned them because their function changes nothing in the logic of the process and to add them to the account would simply prolong the discussion. My purpose is simply to elucidate – and to elucidate simply – what is at work when we observe coloured shadows.

6.4 Looking ahead

To this example, Chapter 7 adds another practical example of an unusual colour phenomenon, that of coloured after-images.

7

After Images

ABSTRACT 7

- After staring intensely at an image and abruptly changing it for a blank page, an after-image appears.
- The after-image reverses the luminosities of the initial image: a black and white image becomes its negative (white where there was black, black where there was white).
- A colour picture appears in its complementary colours.
- The reversal is not related to colour vision. This is evident from the fact that a black and white image also appears reversed in after-images.
- The aspect of after-images that relates to colour phenomena is the change from the initial colour to its complement. This can be understood within the unified principle of colour formation.
- Reversed brightness values yield the complementary colour, as described in the text.
- Although the production of a reversal is not related to colour phenomena, Appendix II proposes a possible cause.

7.1 The brightness reversal

This chapter presents a further illustration of how colour results from a conflict in the brightness levels registered by different cone sets.

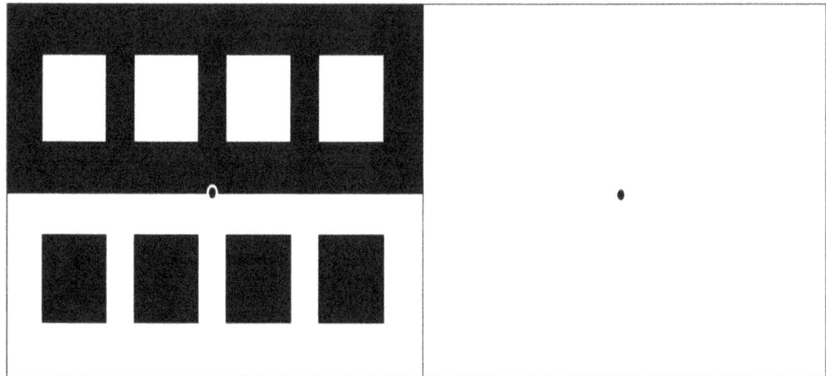

Figure 9. To experience an after-image: Look fixedly at the dot in the centre of the black and white image for some 30 seconds. Then shift your gaze to the dot in the middle of the white space on the right, and watch the after-image appear.

The term *after-image* refers to an image that appears in our mind after having looked intensely at a particular scene. To experience an after-image, stare for some 30 seconds at a picture without moving your eyes and then shift your view to a blank sheet of paper. A moment later, the image reappears on the blank page. The outlines in the after-image are the same as the original picture, however the brightness levels are reversed. What was white now appears black and what was black appears white. If the initial picture was in colour, the after-image is no longer in the same colours, but each coloured surface changes into its complementary colour.

7.2 The effect of the reversal on colour

By recognising that black and white pictures produce a reversed after-image in the same way that coloured pictures do, we can conclude that the reversal is not specific to colour. This means the reversed luminosity of an after-image is not attributable to the phenomena of colour production. The only aspect of after-images that pertains to colour is not why there is a reversal but why the reversal creates the

Figure 10. To experience an after-image: Look fixedly for 30 seconds at the white dot in centre of the coloured squares. Then shift your gaze to the white square on the right and watch the complementary colours appear.

colours it does. The reason for the colour change is elementary and quite straightforward, as presented in the earlier chapters. One paragraph is all that is needed to describe how it works.

In the case of after-images, we know that brightness values are reversed. We know, too, that conflicting brightness values are the source of the colours that appear in the initial image. Taking the example of looking at a blue object: It appears dark to the L cones and bright to the S cones. In the case of reversal, the S cones will register dark instead of bright, the L cones will register bright instead of dark. A conflict that was resolved by seeing the colour blue, now needs to be resolved by using the opposite contradictory ratio. This involves the complementary half-spectrum, a shift from the blue half-spectrum to the red half-spectrum. In this manner, the reversal of brightness relationships always implies the complementary colour, which is why after-images appear in that complementary colour.

If the reader is dissatisfied with this presentation of after-images from the point of view of colour phenomena and wishes to query what brings about the reversal, it should be clear that that is a separate question with no direct bearing on colour phenomena.

To my knowledge, the reason for the reversal in luminosities has not been elucidated to date, although there have been serious attempts at an explanation. The commonly accepted view is that it is caused by eye fatigue, as if the retinal cells active in forming the original picture have tired and are resting, leaving other cells free to become active in their stead. I will not enter this argument, but if the reader is interested in a distinctly different viewpoint, I have included it in Appendix III. I propose that the reversal has little to do with the metabolic functioning of physical cells and is instead a reflex impulse rooted in the perceiver's subjective, albeit non-conscious, anticipation.

7.3 Looking ahead

Previous chapters told how conflicting brightness relationships spur the observing subject to find a resolution through a shift to colour vision. What remains is to understand how the separately formed levels of brightness of each cone set interact with one another and successfully elaborate a common scale of brightness values. The next chapter enters into this.

8

Integrating Brightness Ratios from the Three Cone Sets

ABSTRACT 8

- Each cone set independently establishes its scale of brightness values and they do not share a common unit of measurement.
- Although it is not possible to make a direct comparison of the ratios of brightness formed in each cone set, they can be compared indirectly.
- By comparing 'what is brighter than what' in each cone set, it is possible to form an overall relationship between the different views without requiring units of measure.
- Since the differences between the cone set views result from their three sensitivity ranges, the resulting ratios of brightness will naturally resolve into three contradictory views.
- It would be a highly complex task to work out these relationships in conscious thought, but, when the task is taken out of consciousness, the expanded mind achieves it in an unnoticed moment.

8.1 Three incommensurate black and white views

We return to the problem set aside in Chapter 3, where it was noted that the three cone sets are formed independently (the analogy of the orchestra) and they lack a common measure that would allow for comparison of their brightness values. How then are separate cone cells' ratios of brightness integrated? The broken connection between the cone set views could be described as tragicomic. On one hand, we know that colour can solve any discord, that a potential solution exists and is on standby, so to speak. On the other hand, this solution cannot manifest because of the lack of a common language between the three players – the L, M, and S cone images. Incommunicability holds the views separate, unable to meet and fuse.

Were the views to remain separate, this would upset everything proposed so far. All along, the fundamental premise has been that contradictions in brightness generate the phenomenon of colour. It is here, at the crux of the argument, that we gain a fuller understanding. There is no way to make a *direct* comparison between the brightness in the different cone set views, but an *indirect* path is available. Although the process is complex, with a little imagination it is easy to describe using two different images.

8.2 Resolving incompatible relationships

Example 1

Let us return to the hillside with coloured flowers. Before considering the brightness relationships, note that the positions of everything in the scene are the same for the three cone sets. This invites a superposition of the three views, impeded by the lack of a direct comparison of brightness values (see Chapter 3). The indirect way to achieve a comparison between the different cone sets involves taking account of the *ratios* of brightness, not of the brightness itself.

Here's a description in terms of the coloured flowers. Let's say in one cone set view, the blue gentian flowers are brighter than the green gentian leaves, while in a second cone set, the leaves are brighter than the flowers. This same incongruence between cone sets will be found throughout the scene. Brightness relationships differ in a similar way because S cones register shorter wavelengths as brighter, while M and L cones register them as darker. If we compare 'what is brighter than what' in one view with 'what is brighter than what' in a second view, ratios of brightness can be established without the need to attribute units of brightness to individual items. This establishes an overall hierarchy of brightness relationships without direct comparison.

Example 2

In case this relationship matrix is not obvious, I offer a fanciful parallel from a very different setting.

Two hikers passing through an alpine pasture see a herd of cows. They observe them for a while to see which cow is the leader of the herd, the 'queen' as she is called in the Alps. One hiker concludes it is a certain cow, the other hiker disagrees, having deduced that it is a different cow. In the natural hierarchy of a herd, there can only be one queen and therefore the hikers' conflicting views are contradictory. To resolve the matter, the hikers need to go back and review how they reached their conclusions. If they simply compare the two cows, looking at one and then the other, no feature would mark out the queen of the herd. (I disrespectfully overlook the fact that a seasoned herder might well be capable of doing so). To recognise the queen, it is necessary to observe the relationships among the cows. Only by calibrating the interactions between all the cows of the herd can the leader be ascertained. No direct measure can be made of the supremacy of an individual cow but indirectly, from the hierarchy within the herd, this can be known.

Resolving incompatible views produced by different cone sets presents a similar impasse. Looking directly at an object in the field of view will not reveal its level of brightness. However, information can be gleaned indirectly from the overall scene that enables the disassociated views to be resolved into a single coherent view.

I cannot tell you how the hikers solved the matter but, in the case of vision and colour vision, the creative power of our mind breaks into a novel mode of perception that allows us to engineer a way through the impasse. Such is the potential of our expanded mind. But how and where is this transformation enacted?

8.3 The magnitude of the task of integration

The task of image formation involves integrating the qualities of everything present in the field of view (see Chapter 4). This implies the establishment of relationships between all that is present: spatial relationships (lengths, angles, forms, sizes, etc.) and brightness relationships. In this manner, a separate view forms for each cone set. This process is followed by a more complex stage, the integration of these three independent views that lack a common reference scale of brightness. The monstrous challenge of integrating multiple 'brighter than', 'dimmer than' relationships across the field of view involves simultaneously solving an array of inequality equations. And there is an added requirement: that the whole process is to be accomplished in a split second.

This task appears insurmountable. It seems so because we consider it from the limited perspective of thinking, whereas the task requires moving beyond the one-at-a-time processing of the conceptually conscious mind (Chapter 2). It is accomplished by the expanded mind, where complex processes of integration happen in what, to our conscious mind, is no time at all.

8.4 Complexity resolved by the non-conscious mind

In this process, the three views produced by the cone sets retain their separateness and only produce the colours we see when they are brought together within the expanded mind. The separate black and white renditions take on a new quality and, in a different mode of seeing, are able to form a unified image. To borrow a mathematical metaphor, a sort of Lorentz transformation creates a bridge between views that lack a common measurement scale.

The complexity involved in the resolution of this conflict is not unusual. Quite frequently, tasks that overwhelm the capacity of the conscious mind are dealt with by the expanded mind. Our actions continuously confront the conscious mind with impossible tasks. At which point, the boundaries of what is possible dissolve as we go beyond linear rational thought and let the expanded, nonconscious, multiprocessing mind accomplish the task. As soon as a task is taken out of the grasp of the conscious mind, the possible solutions expand dramatically.

8.5 Looking ahead

Chapter 11 presents some dramatic examples of this uncanny ability. Before that, Chapter 9 proposes a temporal component with the suggestion that the different cone sets do not interact simultaneously, but in a rapid sequence.

9

A Possible Temporal Factor in Colour Formation

ABSTRACT 9

- This chapter explores the possibility that the rate of image formation is different in each cone set.
- Research suggests that the transduction of visual information is more rapid in longer wavelengths than in shorter wavelengths.
- This, in turn, suggests that the L cones process information more rapidly than the M cones and both are ahead of the S cones.
- This would mean that the views produced by the L, M, and S cone sets are formed in a rapid sequence.
- This 'triplet' of images retains its three distinct views on entering the expanded mind. This registers out of consciousness while the conscious mind registers their fusion as a single image.
- The creation of colour through a temporal progression is parallel to the creation of 3D vision in the Pulfrich effect.
- To fuse the incompatible brightness ratios, the perceiver transitions into colour vision (see Chapter 3).
- A cascade of triplets repeated in milliseconds creates a sequence of refreshed images that together give duration to the objects and colours we see.

- Resolution into a single image establishes three things: the perceived object, the perceiver, and conceptual consciousness.
- With an understanding of the unifying process that is the principle of colour formation, we can move beyond the presumption that contradictory brightness impressions function as a coded sequence of pulses. There is no programmed response mapped into our brain that directs the conscious mind to see colour. Colour formation engages the intelligence of a living subject.

9.1 A rapid sequence of views generates the visible image

Three independent views, formed by different cone sets, register in the expanded mind. For them to become the image we see, they need to be integrated into a single, stable, unified image. In this chapter, I propose a sequential process through which this is achieved. It is a staggered process of image formation that also clarifies a number of other situations of colour formation, as pointed out in the chapters to come.

Some research suggests that different wavelengths are processed through the visual system at different rates, the longer wavelengths more rapidly than the shorter wavelengths.[24] If we accept this as the case, it will have an effect on colour formation. Since each cone set is sensitive to a different range (the L cones being sensitive to longer wavelengths and the S cones to shorter ones), we can expect image formation to be staggered. The L cones would process the visual information most rapidly, followed by the M cones, and then by the S cones. With this, the three views form in rapid succession

---

[24] Some work on this (although not linked to what is argued here) can be found in Declan J. McKeefry, Neil R. A. Parry, Ian J. Murray, 'Simple Reaction Times in Color Space: The Influence of Chromaticity, Contrast, and Cone Opponency,' *Investigative Ophthalmology & Visual Science* 44, no. 2 (May 2003): 2267-2276.

and not concurrently. Such a time delay, however brief, would have a powerful effect.[25]

I will refer to the rapid succession of the three cone set views as a 'triplet'. It is intriguing that this visual triplet is experienced not as a onetime occurrence but as a repeated succession of innumerable triplets, as detailed in 9.2.

9.2 Creating colours through a rapid sequence of images

Accepting that the rapid passage of a triplet of images generates a moment of colour vision, then what creates and sustains the steady, constant, coloured image we see?

The answers have already been presented. From Chapter 4, we know that image formation is a crucial stage in visual perception. To become visible, everything that appears in a scene has passed through a process of integration. This process is achieved for each of the three cone sets. We also know that each visible image is short-lived but is refreshed within milliseconds and reforms in a subsequent moment of seeing (see 4.5). In the time it takes to glance at what we see, the image that forms in our mind has been refreshed many times, very possibly triggered by tiny eye movements or by the saccades the eye performs.

The triplet of cone views creates a unified visual image with each refreshed image formation. The sequence plays out over and over again. The coloured image is created and recreated in an imperceptible sequence of triplets that flicker through the mind in milliseconds. To make an analogy with hearing: think of the single sound we hear when a rapid tremolo blends two notes, or think of the note itself, oscillating

---

[25] An example of a dramatic effect produced by a minute delay in processing time is found in the *Pulfrich effect* (see 9.3).

between presence and absence of sound. The expanded mind registers three separate views in a sequence too rapid for consciousness to follow. Our conscious mind perceives a single image from the rapid repetition of triplets, and we see an apparently continuous view of coloured objects.

9.3 The Pulfrich effect

The sequenced effect in image formation that produces colour has a striking parallel in the formation of depth vision.

In most circumstances, 3D vision is a result of the spatial displacement caused by the separation between our eyes, as described in 3.1. What one eye sees is shifted laterally from what the other eye sees. The contradiction in the two views is, to a large extent, corrected when we understand what we see as displacement in a third dimension. In this process, we combine the two images into a single view with depth.

A less well-known way to produce 3D vision is to create two views sequentially, known as the Pulfrich effect.[26] Two conditions are required: 1) that one of our eyes takes more time to form an image than the other; and 2) that what we are looking at (which is in 2D) is moving.

Our visual system functions more rapidly in brighter light. For example, it takes longer to form a visual image through sunglasses. If the view of one eye is darkened, what we see with that eye will take more time to form than the brighter view seen by the other eye. This tiny delay has no observable effect when the scene is without movement. However, the difference can be significant when there is movement and the views of our two eyes are out of step with one another. At the moment of perception, the view from the darkened eye lags behind the

---

[26] A number of videos on the web show this phenomenon: https://www.youtube.com/watch?v=Q-v4LsbFc5c

brighter view formed through the other eye. The darkened view shows an earlier moment, which is shifted slightly from the later view that forms through the other eye, and the result is 3D vision.

In the Pulfrich effect, the displacement is not produced by a spatial difference between our two eyes but by a time delay in image formation.

There is a second possibility to colour production parallel to the Pulfrich effect. We see colour when the three cone sets register contradictions in brightness. Usually, the contradictions arise when the light contains only part of the visual spectrum (see Chapter 3). However, there is also the possibility that full-spectrum light produces a contradiction through a temporal process parallel to the Pulfrich effect and under the same two conditions: 1) that image formation happens at different rates (here, between the three cone sets); and 2) that the object seen (that is colourless) is moving.

This colour-forming process is a kind of Pulfrich effect producing not 3D, but colour. In the case of colour vision, unlike the case of 3D vision, the effect is produced independently by each eye. A detailed description of colour formation by this process is in the supplementary chapter on the Benham top.

9.4 A creative response, not a hidden programme

We return to the three views (triplet) that give rise to colour. As in many examples of resolving a perceptive impasse, the expanded mind accomplishes what the conscious mind is unable to. So, it is with the triplet. Each triplet is a rapid sequence of three differing views. The observer's intent to see gives direction and purpose to the process, which leads to the transformation in seeing that is required for the triplet to appear as a particular colour. Imperceptible eye movements, it is suggested, trigger the image's reformulation and these successive moments of seeing produce the stable, coloured view perceived by the conscious observer.

At this point it would be a mistake to think of the triplet sequence as a flicker pattern that functions as a code and triggers a colour-producing process. This idea could arise from the supposition that the perceiver is responding to a programmed code sequence, implanted in our brain or neuronal system. For colour vision we do not function like a computer, with installed software. Differences in brightness are the origin of colour and the intelligence of the would-be perceiver creates a solution. During the passage of pre-visual information within the expanded mind and its entry into the conscious mind, the perceiver adopts a changed visual attitude – a new mode of seeing. This creative act is an intelligent response to a visual hurdle. The observing subject enacts the change – a change in psyche – that generates the colours we see (see Appendices I and II).

9.5 Looking ahead

The temporal factor presented in this chapter means we need to reconsider the colour wheel as described in Chapter 5. Chapter 10 goes a step further in the discussion of colour relationships.

10

A Second Look at the Binary Structure of Colour

ABSTRACT 10

- The temporal progression proposed in Chapter 9 opens a path to further understanding of the progression of colours in the two half-spectra (Chapter 5).

- One progression is produced in the red half-spectrum, where the three cone set views succeed one another in a transition from darker to brighter.

- The other progression is produced in the blue half-spectrum, where the three cone set views succeed one another in a brighter to darker sequence.

- These two sequences, as described in the text, point to a fundamental structural difference between the two sides of the colour circle.

- In reconsidering the formation of green and magenta, it appears that contradictions in brightness are insufficient to produce these separate, intense colours. We also need to include the contrary progressions of ascending and descending levels of brightness.

- Chapter 5 describes how complementary colours annul colours, but why is the result black, an absence of light? Why isn't it a lighter shade of grey?

- The ascending and descending brightness values annul one another and the presence of light.
- Chapter 10 has proposed some new answers to questions on colour formation. It opens an uncommon approach in which to consider these questions and evidently leaves much for further study.

This chapter takes a deeper look at what was presented in Chapter 5 and goes one step further towards understanding the underlying binary structure of colour.

10.1 The binary structure of colour

Earlier chapters introduced the unified principle of colour that describes colour formation. When the L cones register more light than the S cones, colours of the red half-spectrum appear. At the same time, any brightness registered by the S cones counterbalances the light of the L cones. As the proportion of light reaching the S cones increases in comparison to that of the stronger L cones, the red we see brightens to orange and then to yellow.[27]

A parallel but opposite procedure is found on the blue half-spectrum. When the S cones register a higher intensity than the L cones, any increase in the light stimulating the L cones causes a colour change from violet to light blue. Between the dominance of one or the other half-spectrum, lies a midsection where a balances of the light from each half-spectrum is approached and the combination of brightness values resolves into green (as presented in Chapter 5).

This description presents a first level of understanding. However, we need to delve deeper and reconsider the process that generates the

---

[27] Light from the blue half-spectrum added to the red half-spectrum is not the addition of the colour blue – since blue is the product of a transformation that we, as perceivers, enact. Instead, the entry of light from the blue half-spectrum blends the three different brightness values that tend to be the opposite of those in the red sequence.

two complementary half-spectra. The temporal element introduced in Chapter 9 opens the way for further precision.

10.2 The brightness sequence within the three cone sets

In Chapter 9, the triplet sequence in the cone sets was deduced from the likely fact that longer wavelengths process information more rapidly than shorter wavelengths. I will now describe a second process that determines colour relationships and resides within that temporal sequence. With the assumption that the L cones, more sensitive to longer wavelengths, process light faster than the S cones, we can consider the effect this has on the brightness values within each triplet sequence.

In the case of seeing red, the brightest views are produced first by the L cones and then by the M cones. After these, the third element of the triplet sequence is the image formed by the S cones which, in this case, is the least bright of the three. From this we can conclude that the red half-spectrum forms when the triplet moves in the direction of descending luminosity.

In the case of seeing blue, the brightness sequence is reversed. The first images to form remain those of the L and M cones, but now they are less bright. The view of the S cones, that comes third, is brighter than the other two. When the process within the triplet is of ascending luminosity, colours from the blue half-spectrum are seen.

In other words: descending brightness rules the red half-spectrum; ascending brightness the blue half-spectrum.

When we see a colour, it results from a rapid succession of triplet views that form either in the direction of light to dark or of dark to light. This is a fundamental dichotomy that differentiates the colours of one half-spectrum from those of the other half-spectrum.

It occurs to me that the relationship of the two half-spectra can be represented by the traditional yin and yang image, which portrays the interpenetration of two distinct contrary processes that remain separate.

These parallel but reversed processes present in our visual experience add a new twist to the story of colour formation. The complementarity of two processes yields colour consciousness in all situations and therefore gives a structural basis to the relationship between the blue and red half-spectra.[28]

10.3 A preconceptual visual process

We need to question how this imperceptible sequencing enters our visual experience, given that the progression of light to dark and dark to light is not apparent to our conscious mind. It is too rapid for conceptual consciousness to decipher. The process is a precursor to perception and is, evidently, an activity of the non-conscious, expanded mind. Chapter 11 presents other examples of our subliminal ability to register sensory content that is never consciously perceived and yet impacts upon our perceptive experience. In actual fact, everything we perceive registers initially as imperceptible qualities that are transformed through an act of integration into the object we do perceive. But this is a separate matter and we must not stray from the immediate questions.[29]

A description of how the dark to light and light to dark progressions relate to colour formation are beyond this small book. I will go no further than noting the parallel between ascending and

---

[28] These two reversed sequences echo other phenomena of reversal that are widespread in our observations of nature, whether in plant life, animal forms, DNA structure and maybe even the attribution of left and right spin to subatomic particles.

[29] Further discussion can be found in *Making Sense*, particularly, Chapter 2 (see footnote 3, p. 4).

descending brightness and the binary structure of colour present in the blue and red half-spectra.[30]

It is fascinating that the opposing movements of brightening and darkening must be present in the formation of coloured images we see. These two different progressions surely function to sustain the separation and independence of the two half-spectra. Ascending and descending brightness in colour formation may well be the gateway to a whole new discussion on colour perception, but I will not venture far along this path in the present work.

10.4 The overlap at the ends of the half-spectra

Given this new information, let us reconsider the encounters between the ends of the two half-spectra: yellow with light blue; violet with dark red. When they meet, they produce a new, independent colour (see 5.6 and 5.9). At the meeting point where green forms, the blue half-spectrum wavelengths are increasing and the red half-spectrum wavelengths are decreasing. The progression to green is a dual movement, from above and from below, towards the midrange of the visible spectrum. Where they overlap to form green, the wavelengths will be similar. But if there is little difference between their wavelengths it raises the question: where is the conflict that gives rise to colour? Colour is created to overcome a visual impasse due to conflicting brightness levels between the

---

[30] Discussion of the Benham top is included as a supplement. It is an example of full-spectrum light producing colour, where blue appears in a dark to light triplet and red in a light to dark triplet. A different case of light-dark interaction involving superposition is found in Goethe. He gives examples in which red appears when looking at a light source through a semitransparent medium that darkens the light (thick atmosphere darkening the setting sun); and examples where blue appears when looking at darkness through an illuminated semitransparent medium (the atmosphere in daytime through which we see the dark background of space beyond). An intriguing puzzle for another time is to find the common ground between these different phenomena.

cone sets; however, when little or no conflict is present, there is no impediment to seeing. We must conclude that colour is uncalled for. And yet, where these half-spectral ends encounter one another, we see an intense new colour. What is happening?

The brightness progression in the triplets, described in 10.3, signals the answer. Despite the proximity in wavelengths, the brightness levels of the half-spectra are unable to fuse because they are structurally different. The red half-spectrum is a descending sequence of luminosities, the blue half-spectrum is an ascending sequence. We could say they inhabit different worlds that would demand the oddity of an Escher drawing to bring them into the same space! They remain inherently separate and this is the incompatibility that necessitates the formation of a new colour, distinct from both light blue and yellow.

At the opposite side of the colour wheel, where magenta forms, we find a parallel but opposite situation. Where violet and dark red ends of the half-spectra overlap, the wavelengths diverge; those of the blue half-spectrum become even shorter, while those of the red half-spectrum even longer. Stretching apart to a maximum, moving towards the ultraviolet and infrared, they separate to the limits of what is visible; however, as in the case of green, the brightness differences between the cone sets can be small or non-existent. In all three cone sets, there is a reduction in light. Even so, they do not combine to become colourless (as with complementary colours) but produce an intense new colour, magenta. Surely the result is due to the contrary processes of light to dark and dark to light holding the two half-spectra (violet and dark red) separate. Resolution is only possible with a new colour that unifies these opposing structures of brightness into a single perceptible object.[31]

---

[31] In this extreme case, the L cones register no input from the blue half-spectrum and the S cones register no input from the red half-spectrum.

10.5 Complementary colours produce black

There is more to add about the mixing of complementary colours. As already noted in Chapter 5, the three levels of brightness that register in the three cone sets combine to form a colour from one of the half-spectra. Its complementary colour from the other half-spectra has the reverse levels of brightness. When the two colours are superposed, they annul one another and we see a colourless view. As no contradiction in brightness ratios remains, colour is not present.

So far, this seems logical, but why does the mixture turn black or nearly black? Colourless does not necessarily imply black. The combined luminosities could be seen as grey or white? This is a huge question. Something more than the combination of brightness values is taking place. It is as if someone has switched off the light!

Why does the combination of colours from opposite sides of the colour wheel not simply nullify the need for colour, leaving the illumination as shades of grey? Why does the combination nullify all brightness? The answer seems to lie in the contrary movements within the triplets of each half-spectrum. One sequence (generating red) is a light to dark progression; the other sequence (generating blue) is from dark to light. These triplets seem to cancel out one another and extinguish the registration of brightness. Can we then conclude that, by proceeding in opposite directions, these two sequences neutralise the light that registers on the retinal cells? It is a proposition worth considering.

This last point is where the discussion breaks off temporarily and, as in the tale of Scheherazade, the full unravelling of the story will be suspended. Only, the next episode is more than a day away and, to be truthful, requires deeper consideration.

10.6 Looking ahead

It is my contention that recognising the process of colour formation offers a window into the imperceptible activity of the observer's mind.

The opposite is also true. Recognising the unsuspected abilities of the mind is helpful when considering all that must take place to experience colour. Chapter 11 presents three such cases, from which it is possible to gain a sense of the unobserved processes that enable innumerable factors to fuse and produce a unit meaning that enters our consciousness. The three examples illustrate the presence of an imperceptible mind activity that registers a wealth of sensory information to produce what we perceive.

11

Sensory Discernment
Beyond Consciousness

ABSTRACT 11

- This chapter explores the proposition that colours emerge from an imperceptibly rapid succession of separate views produced by the L, M, and S cone sets.

- It is not possible to observe activity in the expanded mind since it is inaccessible to consciousness, but there is an alternative path.

- Processes that integrate the imperceptible take place in other sensory experiences. By recognising them, it becomes evident that the same can also be present in colour vision.

- Three examples demonstrate this frequently ignored capacity of the mind. They pertain to the ability to gather and integrate imperceptible details (whether they are too small or too short-lived) and to deliver them in a transmuted, perceptible, form to the conscious mind.

- The examples illustrate three types of transformation: temporal-auditory, temporal-visual and spatial-visual.

- How we understand these events depends on whether we think of perception as an automated response to physical events, or as a subject-centred event, set in motion by an anticipation present in a living being.

This chapter shows that processes of sensory discernment, inconceivable to the conscious mind, are a regular part of our experience of being alive. I draw on examples from other areas of sensory experience in which hidden transformations comparable to those that produce colour are common. I would like to think that these examples will encourage us to be audacious and reconsider settled opinions from a novel perspective.

11.1 The transmutation of imperceptible into perceptible

A. A musical note – a temporal-auditory example

If we record the sound of a hammer tapping and then speed up the recording to a point where we can no longer decipher the individual taps, we hear something completely different – a musical note. The rapid tapping, instead of blurring into noise, is delivered to our conscious mind as a new qualitative experience. The tapping has been transmuted into another phenomenon. No longer is there the sequence of noise and silence, instead we hear a musical tone. It might be a low note of less than a 100 beats per second, or a middle range note in the region of 400 beats per second, or a higher note. At these rates, the alternation of sound and silence is no longer distinguishable. But is it really indistinguishable? If we slow the recording ever so slightly, or speed it up a tiny bit, the note becomes flat or sharp. Our conscious mind cannot pick up the hundreds of individual taps per second and yet we are able to distinguish very slight changes that, directly, are imperceptible. Out of the range of our conscious mind, our mind is alert and able to register exactly the number of taps that reach our ears, transmuting them into a new quality that the conscious mind can access. We enter another modality of perceiving. Instead of tapping, we hear a musical note that reflects minute changes that are otherwise imperceptible.

This is an example of a subliminal experience in which a rapid succession of impressions generates a totally new qualitative experience. Our sensory system frequently performs in this manner. It is an example of how an observation directed beyond what the conscious mind is capable of grasping directly, whether in rapidity, size or complexity, does not disappear but transmutes into a different perception of a totally new quality.

B. A film – a temporal-visual example

Let us now consider what happens when we watch a film (a movie). On the screen we watch a succession of individual images that carry incremental changes from one image to the next. When they are projected in fast succession, the conscious mind can no longer follow the individual images. We might expect that this would result in their disappearance or, more likely, in them fusing into an indecipherable blur. This is not what happens. The rapid sequence of images is recorded with great precision in the expanded mind, where it is transmuted into a new qualitative experience now accessible to consciousness: the visual experience of motion.

Even though the images flash past too rapidly for our conscious mind to capture, we do perceive them individually. If this were not the case, we could not observe a clear, well-defined moving image. Our visual acuity is also apparent from the fact that we can sense small changes in the sequences. When images stream past at, say, 30 per second and we increase or decrease the rate by one frame per second, there is a noticeable change. We see the action speed up or slow down. It is evident that we register with great precision small changes in what appears to be imperceptible. The transmutation of still images into movement enables what is imperceptible to manifest in the conscious mind.

Today, we attribute little attention to this transformation. Moving film has become such a common experience in our everyday lives that we take it for granted and overlook the process involved. But I don't think the fascination with visual movement has disappeared. All we need to rekindle it is one of those little booklets of drawings that, when we flick through the pages, becomes animated and surprises us with the sight of movement, for instance, a leopard bounding forward.

These two examples, a musical tone and a film, have a temporal dimension in common. A rapid sequence of static representations exceeds the conscious mind's ability to distinguish between them. But the expanded mind receives the detailed sequence and transforms it into a new modality. This intuits a meaning that can enter the conscious mind and is perceived by the subject as a musical note or as a moving image. These processes would be impossible if they were restricted to the conceptually conscious mind alone.

C. Pointillist painting – a spatial-visual example

This further example of the interplay between conscious and unconscious mind and of the expanded mind's capacity to create a qualitative transformation can be taken from pointillist painting, in which dots of different colours can fuse into an overall colour tone. In Georges Seurat's, *Un dimanche après-midi à l'îsle de la Grande Jatte* we see certain colours when we view the whole painting, but if we look more closely, we see brushstrokes of many colours, including quite different colours from those in the overall impression. A more dramatic example is in some of the paintings of Andrew Wyeth, where coloured surfaces are formed by almost microscopic multicoloured dots. This phenomenon, sometimes referred to as the simultaneous contrast of colours, is well known and well researched and it does not surprise us. But what perceptual process is involved? What is it that resolves the mélange of imperceptible dots into a general hue?

The dots on the canvas are smaller than what we are able to perceive visually from a distance. Our conscious mind does not

Figure 11. A Sunday on La Grande Jatte, Georges Seurat (1884-1886)

recognise the individual dots. They are indecipherable and yet we do decipher them. Were it not so, we would be unable to blend them to produce the hue we do see. This signifies that even though the dots are too small to register in the conscious mind, they are somehow 'seen', that is, they enter the perceiver's act of seeing. The individual dots register in our non-conscious mind, which is where the act of integration takes place, fusing the imperceptible details and generating a new effect, in this case, an effect that expands-over an area large enough to be observable to the conscious mind. What is too minute can then take on the quality required to register as a perception – as the coloured object we see.

Only when we stop and think about such phenomena do we realise how far-reaching a simple example can be. It illustrates how sensory information that cannot be processed in our conscious mind is collected and treated by the subliminal mind. Out of consciousness, our expanded mind transmutes the information into a new qualitative experience. This visual example presents the transformation into a colour of a certain hue and a certain texture that is not apparent in the dots themselves.

11.2 Three realizations

- The conscious mind does not form what it perceives.
- Objects are forged in the expanded mind, out of reach of consciousness.
- From there, what is perceived is delivered 'readymade' to the conscious mind.

11.3 Becoming an object of perception

These three examples illustrate how the experience a new quality through a change in the mode of perception. Such innovation cannot be achieved through a process of construction, that is, by joining parts one to another. Our rational, conscious mind is limited to producing a collage of elements that are quite meaningless until the hidden dimension of the expanded mind intervenes and integrates the parts into a meaningful whole.[32] Only where conceptual consciousness is in 'off' mode, are new meanings, integral wholes, generated. This involves an act of integration that brings a new mode of sensing, allowing for a new unity to form. That newly formed something can then be grasped by our conscious mind as a note, a sense of movement, a colour, and much else in our daily sensory experience. In this way, perception opens up to new possibilities.

The position expounded in this book is that the resolution into visible colours does not emerge from a purely causal process but

---

[32] See *Making Sense*, Chapter 2, p. 71, 'Reviewing Reality' (see footnote 3, p. 4).

is engendered by the aim to reach an outcome. The indispensable requirements for the production of colour are: certain physical conditions, a subject's intent, and the potential for a solution. All three together yield the colours we see.

11.4 Subjectivity

A recognition of the dual functions that are the conceptual mind and the expanded mind depends largely on the role we attribute to the individual observer. It rests on where we place the dividing line between the subjective and the objective. Do we consider the non-conscious activity of the mind to be independent of the subject and determined by purely physical causes or do we consider the non-conscious mind as an expression of a subject becoming a perceiver? We have been trained to accept the former and, typically, to ignore the latter. We may be totally oblivious of the fact that in accepting the first option over the second, we are making a choice[33] and this choice has implications all the way from these few simple examples to the meaning we attribute to existence

----

[33] An obvious challenge to this viewpoint is raised by the question: how can we consider that a personal intervention takes place and a choice is made when the perceiver is not conscious of making any such decision and seemingly has no control in the matter? In this objection, an underlying misconception is at play. It arises from a failure to distinguish between the function of will and that of consciousness. The quality of will that I refer to as intent is not sourced in mental activity, for it functions independently of consciousness. Necessarily, the presence of intent is active *before* the subject is cognisant of it. Only if the living subject gives thought to what it experiences does intent become conscious. In the vast majority of cases, our body and mind respond to the presence of intent without having it become an object of consciousness. For example: when you go to a door with the intent to exit, you slow your pace, lift your arm, position your hand on the door handle and turn it − all this happens without any recourse to thought. It issues from your intent and is achieved without the involvement of conceptual consciousness. There is more to add but, as it is not required in order to understand the colour phenomena being discussed, I leave it here. *Making Sense* contains a more complete description of the source, presence and effect of intent, and of the connection between structures of intent and individual choice (see footnote 3, p. 4).

11.5 Looking ahead

Drawing on what has been presented so far, Chapter 12 describes how we enter the different modes of seeing that express the mental attitudes we adopt.

12

Modes of Perceiving and a Touch of Philosophy

ABSTRACT 12

- Perception is a purposeful activity accomplished by a living subject.
- Sight is not achieved through a self-directed activity within a biological system, it requires a subject.
- The desire to perceive is a deep-seated intent in a living subject without which colour and vision could not come about.
- Vision demands more than looking; it is attitude of mind that structures what we see.
- Visual challenges are surmounted by the intelligence of a would-be perceiver sensing the potential for resolution.
- The potential for resolution seems to call for its enactment – 'could be' attracts 'what comes to be'.
- This suggests a principle of nature wherein entering a new way of seeing is only attempted because the potential solution is present.
- The mode of seeing colour underlying all the situations cited in this book does not exclude the possibility of other modes of colour vision.
- The conflicting brightness ratios of our cone sets that produce a specific colour are the same for different human observers. We all have the same conflict to resolve.

- But, is the experience you call *blue,* the same experience I refer to as *blue?*
- 'Yes, it is!' is the apparent answer because:
    1. You enact the same unique resolution that I do because we perform the same integration that unifies the brightness levels of the L, S, and M cones.
    2. The three levels of grey only become visible when transformed into a unique hue.
    3. That hue and no other enables them to be co-present.
- Therefore, surely, we share the same experience of colour.

An understanding of colour vision and its achievement by a living observer can be deduced in clear, logical steps from observable phenomena. It was this realisation that gave me the impetus to embark on the study of colour that led to these pages.

12.1 Modes of seeing

To return to the central question: when we see coloured objects, what enables the resolution of the three incongruent brightness levels produced by the different cone sets? What exactly is our individual role in the act of colour perception? The answer may be a surprise, even though it is quite evident.

A resolution is possible because seeing involves more than simply looking – looking is not enough! Seeing requires that the eye functions in tandem with the mind, that is to say, in tandem with a mental attitude that determines the mode of seeing the observer employs. There is more than one mode and a variety of mental attitudes that can activate seeing. Chapter 3 looked at how depth vision comes into being through a mental attitude that is different to that which recognises a 2D surface projection. Seeing a black and white image is a mode of seeing that only registers brightness levels. It cannot cope with simultaneous variants and can do nothing with the conflict that arises when the illumination reaching the eye is not full-spectrum light.[34] This mode cannot go beyond these differences to attain the unification of the three views.

The intent to see leads the would-be perceiver beyond the limits of this mode of seeing and through the obstruction into another mode that can integrate the contradictions. Just as a runner continues moving forward at a hurdle by jumping, so the visual observer has to make a new move to continue towards the goal.[35] The new move is to accept simultaneous combinations of contradictory brightness levels and to brave the contest between them to allow the transition into colour vision. I speak humorously of it being a contest. In fact, the hurdle dissolves before the subject's intent, the contesting possibilities give way naturally to the new vision of colour.

But we need not be fooled into thinking that this happens 'automatically', as how things function on their own. It happens because the observer is an intelligent being that registers the impasse and invents a way to move beyond the impediment. An observer, without making a conscious choice, gathers the conflicting views into cohabitation. This achievement, it is true, is possible because of the nature of what is there: the initial conflicting values hide the conduit to a solution that allows *seeing* to become *perceiving.* But the conditions do not produce the act of seeing; this requires a living, intelligent subject whether a human being or tiny insect being.

Even though colour has been generated by the living subject's mental attitude to the situation, the observed colours appear as objective facts that enter the perceiver's consciousness.

---

[34] Appendix I expands on this thought.

[35] As an analogy, the hurdle is only useful to illustrate the change in movement. Having jumped the hurdle, a runner returns to running on the ground while the observer, having jumped to a new mode of seeing need not fall back to the previous mode.

12.2 Considering the implications

To claim colour vision as 'a mode of seeing' has implications beyond our present subject. Conflicting brightness levels guide the visual process into a colour mode of seeing, but they only guide the observer to a change in consciousness, they do not create the result. For this, an act of mind is required.[36] This is significant. The observing subject instigates and undergoes the change of seeing that allows colour to emerge.

What I have proposed does not exclude the possibility that other ways exist for a subject to enter the mode of colour vision. To me, however, no such other alternative is apparent.

The subject's involvement in forming the perceived object is essential, even though that involvement does not register in the subject's consciousness. Structures introduced by the perceiver are necessary for the perceived object to take form. Such structures are non-physical presences, one might call them psychic presences, that have a definite role but are not reflected upon consciously. They are non-material anchors necessary for recognition of anything. Without these attitudes of mind, no object can be perceived (see Appendix II).

12.3 The point of view adopted

Chapter 11 gave several examples of the ability of the expanded mind to process an uncanny amount of material in what seems to be no time at all, and to generate a unified perceptible outcome. The common view in our society is that this is an autonomous activity by millions of cells in self-directed metabolic processes. The approach taken in this exposé, in contrast, is to view it as the activity of a subject

---

[36] Further remarks on the instrumental role of the mind in sense perception are noted in, 'To see without eyes; to hear without ears' in Chapter 2 of *Making Sense* (see footnote 3, p. 4).

generating the process of perception so as to achieve meaning and to experience an intelligent interaction with its environment. Describing it this way is a minor change from how it is generally understood, but the implications are major.[37]

It signifies that the process of perception, both non-conscious and conscious, is generated by a specific intent held by a subject. Molecules and more are not simply moving about following the energy patterns that draw them here and there, but there is a reason that all flows towards a goal. It involves purposeful action aimed at a specific achievement. Perception is not self-generating; it is the expression of a being in a first-person position with an intended end. A subject generates this activity through a specific desire or intent, in this case, to see something. This motive force can be understood as the expression of a wider purpose contained in an individual's life mission. It is the reason a being has for living, not for survival, for hanging-on, but its reason to be alive.

This is the point of view I have taken and is why words describing purpose appear in this book, despite being shunned in most scientific discourse.

12.4 Is your blue the same blue as my blue?

There is an age-old question that asks, is the colour you experience when you say *blue* the same visual experience I refer to as *blue?* This is not the anthropological question of whether different cultures experience colours differently, but simply whether blue, red, yellow, and so on are the same for both of us.

Each viewer creates her-his-its individual solution to a visual impasse by generating the colour required to see. Only that specific colour has the capacity to blend the three contradictory renditions

---

[37] See the Epilogue for further commentary on a subjective point of view.

of grey into a single visual experience. In any given situation, you and I face the identical difficulty. The lighting is the same for both of us and, as humans, the sensitivity ranges of our three cone cells are basically the same. One colour, and one colour alone, can overcome the conflicting views. And so, if you will permit, I propose that, 'Yes, our experience of colour is shared!'

To confirm this, I would need to see through your eyes, just as you would need to see what my eyes and mind produce. But still, the understanding in these chapters about how and why a specific colour is produced, points to the fact that we face the same conflict in brightness levels, and resolve the dilemma with the only hue that can balance the different intensities into a single, unified appearance. We vanquished the same visual impediment by employing the same method. Surely, that makes it more than probable that our experience of colour is the same!

12.5 On how the world works: a digression for the curious

I believe that the process of bringing together conflicting views happens because a solution is possible, and only because of this. Such a thought is not unknown in both physics and biology. The process that reveals colour does not involve trial and error, rather, it exemplifies a process enacted specifically because the solution is potentially present.[38]

The achievement of colour vision is not a chance outcome in a stochastic series of events. The process is not arbitrary. The movement

---

[38] The implication here is that if no solution were possible, the three cone views would not be brought together into conflict and there would be no attempt to produce a solution. In that case, consciousness would not appear. This principle has a larger place in my earlier work *Making Sense,* p. 173, 'A View of Evolution' (see footnote 3, p. 4).

towards this outcome is set by the fact that specific colours hold the potential to resolve specific contradictions inherent in colourless views. Much more than simply *hold*, colour *is* the potential being enacted.

Could it be that 'the potential to resolve' actually functions imperceptibly as a behind-the-scenes influence that induces the phenomenon of colour vision? What does it imply if, in our world, 'could be' exerts an influence on 'what comes to be'? To posit such an influence on events raises questions that lead beyond the subject of colour vision and beyond what is included in this short study.

But stay yet a moment. We can also ask whether this inductive effect is produced exclusively when the subject is a living being? Is it also present in inanimate nature? Might it be an unexplained dynamic property of the physical world? This is not a trivial question. For the present, I say simply, 'to be continued' – but elsewhere.

13

The Perceiver Wields the Key that Unlocks Colour

Having traced the stages in colour formation, we can summarise the process as follows. When three incompatible views enter our preconscious mind, they become an impediment to seeing. This obstruction contests our instinctual intent to see and recognise a visual object. As a living being, we take this as a challenge, a gauntlet flung before us. Faced with the impossible, we break into a new way of seeing that unifies what was divergent. But where does this transformation happen? *Nothing changes* in the perceiver's physical context; *nothing changes* in the luminous influx to the eyes. Incompatible shades of grey register on the retinal cones both before and after the event and yet, all has been transformed. A shift takes place within the observer's attitude, it is there that the change in orientation is enacted. The observer, becoming perceiver, welcomes the contradiction. Conflict changes into cooperation. Differences in brightness ratios merge to become participants in the production of a new way of seeing. Thus, the would-be perceiver gives birth to a novel, unified experience – the manifestation of colour.

With this understanding, I can conclude the presentation of the unified colour principle. The essential stages of colour formation have been traced out. Related considerations are in the Epilogue and three Appendices.

Although the principal story has been set out and despite delving no further into what has been proposed, I am not yet ready to let go. The final part of this book is a separate, rather lengthy, study on a well-known enigma: the appearance of colour on a moving black and white pattern. It is a work in progress. It is the case of the Benham top. When the black and white top is illuminated by full-spectrum light, no brightness variation results from different wavelength sensitivities of the cones and yet, colour is produced. As a puzzle with no clearcut answer, it is an inviting challenge for the unified principle of colour.

14
Epilogue

A living being,
    permeated with the intent to see,
      possessing the intelligence to comprehend,
        transforms sensory stimuli into the object it perceives.

14.1 A non-physical influence

The description of the process of colour formation in Part I of this book proposes a changed vantage point for describing how a living being – body and mind – attains perception. It includes a step-by-step presentation of the physical and biological factors that lead to colour vision, but its significance essentially lies elsewhere. It resides in the shift out of a third-person account of physical interactions and into a viewpoint that includes the effect of an intervention by a living being. Some factors, like the desire to see something and the ability to recognise an object, only become operational when an individual is present and realises a perception. These factors sustain the engagement of the subject and can be described as the motives that animate the being becoming perceiver. Beyond cell interactions in the presence of light, non-material factors generate and guide the accomplishment of perception by a living organism.

This accomplishment requires 'teamwork'; the collaboration of an uncountable number of interactions within the sensory system. In the frame of reference proposed here, these are seen not as individual causal effects but as the expression of a shared, goal-directed activity. The common purpose is rooted in a living subject. It is the subject that gives meaning to the act of perception. In fact, without the subject, what meaning is there in the word 'perception'? We have become accustomed to a phlegmatic view of the process, wherein the subject is seen as a passive onlooker while the biological activity proceeds on its own. This view limits the subject's activity to the position of witness. I propose a very different role for the subject, that of a subject-creator determining its own reality.

As a subject-creator, perception issues from the subject's engagement with an outcome. It requires binding sensory impulses together to create an intelligible object. The subject's engagement can be understood as the intent to perceive.[39] Every single thing we recognise emerges from this commitment on the part of the would-be perceiver. The intent of the individual subject has the capacity to transform sensory impulses into information and therefore the subject is inevitably engaged in the creation of what is perceived.

As perceiver, the subject is an active participant, injecting direction into all that comes to pass. Chapter 4 shows how achieving integration across the field of view requires the subject's intelligence. Chapter 3 describes the necessary, intelligent, intervention of the would-be perceiver when faced with unseen, incompatible cone set views that obstruct the process of seeing and generate a need for

---

[39] In this work, the word *intent* is employed according to its common meaning: a subject's commitment to a specific outcome, to what is intended. This is not the same sense that is attributed to the term *intentionality* in phenomenology. In *Making Sense*, I show that intent is not sourced in the mind, not in consciousness nor in thought, but that it has a separate origin (see footnote 3, p. 4).

resolution.[40] Such a process can no longer be described in terms of cause and effect interactions. It is dependent on factors that have no physical presence and yet are instrumental to accomplishing this perception. These factors issue from the subject's living experience and from its freedom to act in a particular manner.[41] The source of this activity is the intent of a living subject. To actually see a visible object calls for something more than the energy exchanged in biological processes; it alludes to an active influence of a different nature that has the capacity to transform sensory stimuli into meaningful content. A shift in consciousness is required, a redirection of the observer's aims. Instead of merely transfer and transformation of signals, there is a recognition of meaning, of the message-content carried by the signal. Perception occurs when the intelligence of the subject enables the transmutation of impulses into meaningful content. Stimuli that impinge on our sensory organs sustain a shared purpose that manifests when the subject recognises a particular thing, an object of consciousness. The triumph of perception is the realisation of an intelligible presence by a subject.

14.2 What the scientific method omits

The scientist or person who believes in an independent objective world, accessed by sensory experience, may find no place for these considerations. The absence of physical cause and effect sequences in perception leaves such a person without a criterion for what is real. This material-based attitude is widely adopted and is generally accepted despite the fact that the postulate that claims the physical

---

[40] This was shown to be the unified principle central to colour formation: the production of incompatible views that require a resolution through the creative initiative of the would-be perceiver.

[41] That intent and choice are disengaged from thought is elaborated throughout *Making Sense*, in particular Chapter 9, 'Freedom of Choice' (see footnote 3, p. 4).

world is autonomous and independent of the perceiver has never been demonstrated. The material-based view is a building with only upper floors that was never set on the ground. What manifests to our senses is accepted as independent reality. Everything that precedes an object's perception and leads to the recognition of its physical presence has been ignored.[42] Somehow, we manage to overlook the activity of the individual being that is imbued with the intent to generate sense and to reveal meaning. And yet, without that presence sustaining this intent, there would be nothing more than disturbances and meaningless impulses impinging on the sensory system. Without the integration achieved through the subject's desire to make sense, nothing sensible or perceptible would form.

In taking account of the direction-giving influence of intent – that of purpose and with it, the engagement of the subject's intelligence in reaching an outcome, we discover that the causal processes we know from the study of physics cannot determine what appears visually to the observer. Something other than the material cause-effect relationship is involved. In its stead we find a process in which the goal influences the means, a process in which anticipating a particular outcome creates the tendencies that lead to its manifestation. While the scientific approach begins with the perceived object, the view expressed here takes account of that which precedes its recognition and enables the object's entry into consciousness. Physics still has its place and causes still generate effects but, in acts of perception, there is much yet to recognise that is not in the purview of physics.

---

[42] This does not lead to theoretical assumptions about the origin of matter and its submicroscopic constituents. It is not matter that is being questioned, but the process of perception.

14.3 Material and non-material presence

Non-material presences such as form, beauty and colour are not to be found in the constituent elements that render these observations. Colour, as has been shown, is not present in the light but is an achievement of the perceiver, and beauty is not found in any one element but in an appreciation of the overall composition.

We do not find the beauty of a painting in one element, such as a brushstroke, but by grasping the whole composition at once; and the beauty of a brushstroke is in its overall form, in its unfurling, and not in any one part of it. Beauty is only possible in all-together-at-once. The unity and the presence of a whole, of a something – whatever it is – comes from being all it is.

Unlike physical objects we know through our senses and of which we typically ask, 'What is it made of?', non-physical objects, and this includes all concepts, require that we encompass their 'all-together-at-onceness' whenever they enter our thought. It is as if the object we recognise is drawn from myriad imperceptible sources that fuse to become that specific thing which crystallises in and as a thought.[43] (A simple version of this process is described in 4.2 'Forming a visual object'.)

From this wider viewpoint, we can recognise two opposing processes through which we know what we know. One process, (a), pertains to non-material presences, of which beauty is an example. The process begins imperceptibly, at the periphery of what

---

[43] This fact is elaborated in the book *Making Sense*, where it is shown that anything we recognise is what it is, not because of some specific content or attribute it has but through a distinct process that requires an act of integration. The object we recognise, every single thing we know, is formed from the outside-in, where multiple impressions come together in the expanded mind and fuse into a recognisable unit that manifests as the perceived object. This 'coming together' yields consciousness and far more than we might first imagine. For example, see the video at *www.petermoddel.ch* titled: *What can matter be?* It illustrates the fact that matter is formed through the same process.

is recognisable, and ends with a unit outcome – a perceived quality or an object itself. The other process, (b), its complement, pertains to observations of the material world. This understanding begins with a specific, such as a physical object which produces an effect that expands into the physical world, dissipating ever more widely. The first, (a), is a centrifugal process that concludes with a unitary object; while the second, (b), is a centripetal process that begins with a recognised unit.

Process (b) is that of a billiard ball knocking another, producing further movement that is distributed every more widely and dissipates finally as warmth. This is the process we know as causality. Process (a) yields all we know through the power of the mind or, using the term proposed in Chapter 2, all that registers in *conceptual consciousness*. Every observation we make is an example of this, such as recognising a fragrance, a flavour, a sound, a feeling, a sight, a sensation – whether they are attributed to the world or as something that arises purely in the mind.

Which of these two processes we recognise depends on our point of view. It is as if we are perched high on a mountain ridge that divides two watersheds. I'm thinking of the Rocky Mountains, where rain that falls on one side will eventually find its way to the Atlantic Ocean, while rainfall on the other side flows to the Pacific. From this vantage point, the side we look at determines our perception of the result. In contemporary thought, it seems to me, we pay more attention to side (b), the side that presents causal processes. Its overall movement is dissipative. It manifests as an expansion and loosening of limited structures and this is referred to as entropy. The other view, (a), is a generative process that manifests wherever we recognise life. It also includes the mind processes that produce intelligibility. In a world charged with intent, the generative movement is expressed in all living things and, of course, in ourselves, our choices and our ability to experience the presence of 'otherness' – presences separate from our own. Process (a) is that of intent; process (b) that of causality.

Rather than a finding a balance between the two views, we've been captured by the view to one side, to that of causal processes. In earlier times people marvelled at creation, at its interlocking processes and at the intricacies found in everything that nature manifests. This was expressed in many ways, such as the creation of artefacts invariably enhanced with artwork. Naked functionality was not felt to be an end in itself; something more was required. All of life nurtured a sense of the beyond and the belief that there is mystery in all we observe. This attitude was generally accepted, in fact, it was so obvious that theologians employed the marvels of nature as an argument to prove the existence of God. Since then, microscopes, telescopes and much else have revealed a complexity and interdependence far beyond what was previously known and yet, surprisingly (allowing for some exceptions), our appreciation of the marvellous has not grown. In fact, it is the opposite in many areas.

In specialised scientific research, we frequently come upon an indifference to the wondrous expressions of life. In this context, to point to an apparent presence of purpose in events and to a direction in natural processes is to introduce something awkward, something suspect that can be seen as a lack of scientific rigour. Qualities and a sense of purpose prove problematic because they cannot be defined in the same terms as material events. In the search for knowledge, an admiration of the marvellous has been relegated to a diminished position. It is accepted as an aesthetic quality, a tangential byproduct, non-essential to its existence. There is a reason for this. There is no place for achievement in a world where all is assumed to be driven from behind, that is, by causes that produce unintended results. Achievement, in a world of random events, makes no sense! The flower then, through chance events of evolution, has a particular role in a causal chain that results in the production of seeds. In this approach, other perspectives are distanced. The flower does not bloom for the bee and most certainly not to delight you and me. It is defined instead by its function in the plant's reproductive cycle.

Intent is free from such assembly-line restrictions. Intent draws from multiplicity, from the unlimited and even the unfathomable, so as to generate an outcome that can be perceived and experienced. To frame the same process in other terms we could say: intent is present in activities that are the response to an attractor. That attractor is the anticipated outcome – the outcome anticipated by the subject.[44] It is produced from within the subject, a subjective quality, and not from an outside influence.

As an attractor, intent is active in everything we see, sense, recognise, and know. Even so, this fact is generally ignored. We are part of a culture and an intellectual milieu that belittles and ignores the presence of intent. I believe the purposeful activity of intent used to be more widely recognised as a dynamic and generative influence in the world, but this has diminished over the past centuries. This gradual loss is inversely parallel to the growth of scientific thought. Since the Renaissance, this formative principle has receded and blindness to intent has left its mark on contemporary culture and on its influence in the world. We readily acknowledge disintegration and loss of structure, the (b) view above, without recognising the complementary generative force of (a), so fundamental to life. Although it is accepted that life processes taken in isolation manifest the inverse of entropy, the importance of this is masked when life processes are understood as a

---

[44] In intent, the attraction to a particular result is due solely to the presence of a subject, the individual being that senses anticipation. Note that the subject's anticipation, as expressed here, is not transferable to the concept of teleology, which refers to an end that is programmed into events and taken, for instance, as a fact of nature. This is not at all the same as anchoring the anticipated outcome to the presence of the intelligence and the intent of a living subject.

subset of an inanimate autonomous world, responding and subservient to material causes.[45]

In a world lacking purpose, that is the result of random events, meaningful outcomes are described as statistical flukes. Where cause and effect processes rule our thinking, qualities have no intrinsic place. It is likely that this results when the (b) side of the divide engulfs our thinking and overruns the (a) side that brings together understandings from the expanse beyond rational thought.

14.4 In a purposeful world

Surely it is not surprising that the experience of sight is animated by the anticipation of seeing. From this viewpoint, the details of cell interactions become almost irrelevant because they go no further than pointing to the 'machinery' employed by the intent to see. These 'working parts' cannot recognise the impasse to seeing, nor sense the need to achieve a coherent unified image, nor have the inventiveness required to find a solution. These abilities flow naturally from the subject's intent. They require agility and intelligence and are sensed on a deeper level than conceptual consciousness. Sensing purpose is a property of the would-be perceiver that generates the diverse solutions needed for vision, such as the act of integration, the sense of three-dimensionality, and of colour formation. The intent to see and to create an intelligible visual object cannot be directly observed and is necessarily missing from a view restricted to material exchanges and their related forces.

Flowers full of succulent nectar and bees with the capacity for foraging do not tell us how honey arrives in the hive, never mind how

---

[45] Life, when viewed as a separate activity, does not function in accordance with entropy. And yet, the uniqueness of life is explained away when it is seen in terms of the overall exchange of entropy in the universe. Then, inserted in the wider view, its special quality can be reduced to thermodynamic principles and it is not recognised as the achievement it is, as a generative force.

the interlocking functions of the natural world generate the living forms we observe around us. There is something more that is not found in the elements of the scene and it is the presence of intent, harboured by individual bees and manifest in honey. Intent is apparent in the result. It might also be perceptible in the way different activities move towards a common goal. For example, the petals of certain flowers curve in a way that augments acoustic resonances specific to the bees that visit them. How much do we really know about how the world functions?

This anecdote about honey points to the necessary presence of a subject – the bee or me – to reveal what we know as the world. The individual subject is the focal point that anchors undefined presences and realises them. To dismiss the role of a personal subject and reduce subjective experience to an objective account is, in my eyes, disappointing. It is the loss of the impulse that animates life: meaning, beauty, joy, purpose, and the mystery that gathers all-that-is into our consciousness. How strange that science imagined it was possible to fully describe life without including the subject's engagement, for surely the engagement by the individual scientist was the impulse for the scientific research in the first place!

The subject's desire to see generates ways of achieving the intended goal and does so by employing the biological structures that have been gifted, seemingly, for this purpose. Whether we refer to the special quality of engagement as anticipation or desire or intent (the last suggests a deeper anchor), it has a role in achieving outcomes. The process involved in colour vision is a powerful example of the necessity to include subjective involvement and recognise the presence of an active subject engaged in producing that which becomes the perceived object. The creation of colour is an illustration of a generative life process that is imperceptible in material interactions, but very much present when we trace out the whole process. Under a microscope nothing is revealed; individual parts of the process divulge nothing. But when we grasp the overall movement, we can deduce the imperceptible correlation leading to a visible outcome – in this case, the emergence of colour – and with this, acknowledge the presence of intent. From an observing subject imbued with intent, an object is perceived. Objectivity is formed from subjectivity. We are individual explorers and individual creators, bringing insight and understanding to our experience in and of the world. From our individual vantage points, each sustaining the intent to see something, we clothe experience with recognisable qualities and dress it in objectivity.

To close, I would like to share a personal anecdote. Understanding what was missing in our vision of the world, and recognising how to reintroduce it became a driving force in my life. Equipped with a pile of empty notebooks, I lived in a cabin in the Himalayas for 12 years (with interruptions) preparing the book *Making Sense* (see footnote 3, p. 4). I realise now that I wanted to compress more than I was able into its 250 pages and I am grateful to have the opportunity to revisit some of those earlier themes in this small book that focusses on just one subject, the presence of colour.

Appendix I
The Chameleon
and Our Sense of Self

This appendix relates to Chapter 1.7.

We are limited by our conscious mind because it is ruled by linear, sequential processes. To see what this means, let us take on another perspective and, for a little fun, try and imagine how a chameleon would cope with the limitation of only one view at time, as we do?

The chameleon's two eyes function independently and each one can look in a different direction. We could compare it to looking in a rearview mirror while cycling, but it is not the same. We switch between forward and backward views, but what would it take to actually see both views at once? We cannot do it using our normal consciousness. It would be like listening to two different stories or two different melodies at the same time. We'd feel as though we'd split ourselves in two, and had two life stories unfurling in our conscious mind at once. As long as we hold on to our unique 'I', it won't be possible. If the chameleon made sense of the world as we do, it would feel incapacitated.

However, it might be a possible to enter two realities if I could step back from myself and did not conceive of myself as an agent acting out my life in the world, but rather sensed myself as a function of all that is going on around me. Then the entourage would be part of me and, like a hologram, I would be the expression of all that is happening. But I am not a reflection. I am a living being responding to events. Yet, to continue the thought, in a holographic state my identity would dissipate as I change in relation to events around me. I can imagine being able to intervene in some and move away from others, while all the time being 'in the flow' as an expression of all the interactions taking place. I could have any number of eyes, looking in any number of directions, far and near, and wouldn't be confounded by my sensory experiences.

Such an all-inclusive relationship comes from a loss of separation. I no longer consider myself a separate entity, with me as the glorious fool in the midst of a whirlwind reality. When I cease to separate myself from all that is, I am no longer restricted to one-at-a-time processing, to being the one experiencing the many. I can be many while functioning as a presence, an entity, among all.

The chameleon matches its colour to its surroundings, which suggests, symbolically at least, a blending with all. Other animals share this ability of being part of all, and some, perhaps insects, can experience many different views at one time.

Is it possible that, for us, the three separate views produced by three cone sets are not in conflict when this mode of understanding is reached? The differences can coexist in the expanded mind where they are not troubled by the formation of concepts, and the observer is free of being distinct from what is observed. As soon as concepts form, so does the relation of a subject separate from an object. This does not happen in the expanded, non-conscious, subliminal mind. There, distinct, uncompromising views coexist and generate a new way of being out of the co-presence of different cone set views. The contrasting views are the invitation to which the observer, as volunteer, responds, taking up the challenge. Instead of seeing contrasting bright-ness values of darker and lighter, the differences are brought together. They manifest as blue, green, yellow or pink, or to put it differently, as cheerful, doleful, aggressive, sweet, calm, lively and all that enriches the experience of being alive. In seeing colour, we identify with its particular experience.

What I've just said is not really about the colours that form, it's about the mind of the perceiver during the transmutation of black and white into colour. It is about losing the sense of self during a timeless moment and transitioning into a new mode of seeing. It is about the generation of a new attitude while absent in the imperceptible expanded mind, free from the sense of self.[46]

What I've just said is not really about a chameleon either. That wonderful creature has shown us a way to loosen our conceptual thinking, to step out of time and make our separateness disappear while multiple views coalesce. Then, in the next instant, to come back to our self as we acknowledge what we see before us, the magnificent blue, the translucent yellow, and all that constitutes our world.

---

[46] See also *Making Sense*, Chapter 7, p. 186, 'The Disappearing Self'; also Chapter 2, p. 51, 'The Disappearing Object' (see footnote 3, p. 4).

Appendix II
The Undeclared Power of Perception

This appendix relates to 4.2 and 4.3 and wherever perception is discussed.

Perception requires the active intervention of a subject. The following examples describe the perceiver's role in greater depth than before.

Consider the case of seeing motion on a screen. It seems quite obvious that a rapid sequence of images, depicting the displacement of an object, produces what we recognise as motion.[47] But this cannot happen without very specific input contributed by the observer. What are these unrecorded additions the subject introduces?

To see motion, there must be an understanding of continuity. To integrate a sequence of different positions and to experience them as movement, the observer must have a sense of spatial continuity and must assume its presence. Without recognising a continuous spatial extension, the changes cannot be understood as the object's

____

[47] Here I speak of the conceptual understanding of motion, our thoughts about motion. The *experience* of motion is sensed through the body and this, in all probability, is a prerequisite to our *conceptual understanding* of motion.

displacement. A sense of continuity also requires a grasp of a temporal dimension, that is, of continuity in time. Bound up with this is the understanding that the moving object is identical with itself during the process. All this is required in order to decipher what we observe.

The source of these understandings is in the subject's act of perception. By applying an understanding to sensory impressions, we integrate diverse stimuli and create the meaning we experience as perception. We have an intimate relationship with what we recognise, for we give rise to the presence of all we know and accept as our world. Without our subjective contribution, we would be unable to perceive motion, objects, events, and all we elaborate as the world and ourselves. What mighty significance this realisation brings!

But let us stay within the immediate context. The observer needs to apply an understanding of the continuities of time and space to sensory impressions in order to perceive, for example, a row of lights switching on and off in rapid succession as a single light moving through space. The application of specific conceptual structures renders sensory stimuli intelligible. These anticipatory structures are introduced by the observer and form what is perceived for instance, perceiving an effect as an object moving in space.

Our culture has a deleterious habit of discounting personal intervention.[48] Perception is attributed entirely to activity in an external, objective world. Our body and our sensory system are understood as part of such an objective world. By extension, this view leads to the unfounded belief that continuity is a feature of the world and not a mode of perception.

In my understanding, the same misappropriation happens in the case of colour, when we imagine colour as something objective, belonging to

---

[48] To free ourselves from this habit we need, among other things, to recognise that personal intervention does not require thought (see footnote 33, p. 95).

the material world. In attributing the phenomenon of colour to objective reality, we overlook the necessary active intervention of the perceiver. We overlook the fact that brightness, distance, movement, and more are qualities we generate personally to perceive what we recognise as the world.

Several further examples of subjective intervention can make this fact more obvious.

What determines whether a series of notes do, or do not, form a melody? The answer is not found in the notes themselves. It is the perceiver's ability to integrate them into a coherent expression. Similarly a chord is a different entity to a note, not because of the extra notes it contains but because it has a quality of its own. The same can be said of the different qualities of major versus minor chords. A more striking example is the seven Greek modes for they are formed by the very same notes and yet each mode is a distinctly different expression.[49] The difference lies in the mind structures we, the listeners, fit on the sounds and thus enable a perception of what is heard.

To take a final example from a different realm: when you experience a sensation of touch travelling down your arm, what determines that it is a caress? We may imagine this meaning to be something we add to the physical event and, true, it is that but no more so than the understanding we bring to every experience in order to transform sensation into meaning.

We integrate sensations and experience their presence as a recognisable quality with a particular meaning. This is what I refer to in the Epilogue as *meaning*. The stimuli that enter our visual system are unformed and unintelligible until, as perceiving subjects, we enact the transformation into meaning. Colour is produced in this manner.

---

[49] The seven modes, or scales, can be played using only the white notes on a keyboard. Each begins on a different note: Dorian on D, Phrygian on E, Lydian on F, Mixolydian on G, Aeolian on A, Locrian on B, and Ionian on C.

The integration yields an understanding that goes beyond what the elements that constitute that understanding convey on their own. The whole is not the sum of its parts. The mind moves out beyond individual impulses to transform them into a new modality. Perception expands to include meanings not contained in the sensory units.

Perception cannot be described as metabolic processes sustained by an independent being sensing an independent world – as if looking were an act of gathering what is already there. Each moment that brings intelligence of the presence of something is, in fact, an intervention by a perceiving subject integrating sensory input and creating a specific understanding. To follow this realisation to its full implications is an adventure in itself; however, our goal is not a philosophical exploration but a description of the phenomenon we know as colour. I wish to demonstrate how, through subjective acts of understanding, we generate what we perceive to be colour and, by implication, erect our objective world.

It might seem odd to think of ourselves as creatures perceiving a world which we ourselves engender. In this, are we not somehow similar to the snail or the tortoise that forms the home it carries wherever it goes? We generate the world that cradles us. We elaborate the world that welcomes us. Both guest and host in our intimate world, we partake in the unfathomable presence we call the universe.

Appendix III
Luminosity Reversal in After-Images

This appendix relates to Chapter 7. Its purpose is twofold: first, to open up a new direction of thought concerning the phenomenon of reversal in after-images; and second, to further the argument supporting the necessary presence of a non-physical influence on the achievement of perception. This influence is present in the attitude adopted by the observer.

After-images: a quick reminder

After an image is stared at, then removed and replaced by a blank page, an after-image appears, seemingly projected on to the blank page. It reproduces the original image but the luminosities are reversed, dark to light, light to dark. Any colour in the initial image changes into its complementary colour, for instance, yellow changes to a blue-violet.

Entering the question

The reversal in after-images is independent of colour formation, as is evident from the fact that the very same change between opposites takes place in a black and white image (see Chapter 6). What is it that produces the reversal in brightness values? Although this question does not concern colour vision, it is revealing and I take up the challenge to propose an approach.

Troubling facts frustrate an explanation. In the reversal, unstimulated cone cells in the initial image begin to respond as if they were illuminated. This is strange. It is also strange that cone cells which registered the bright parts of the first image suddenly respond as if receiving light of the corresponding but opposite intensity. When colour is seen, the different brightness values of the three cone sets are exactly reversed, as shown by the fact that a coloured surface becomes its complementary colour (see Chapter 6). For these reasons and more, to interpret the reversal as the result of eye fatigue is not the answer.

Rather than trying to solve the puzzle on the level of physical stimuli, I suggest we look at the subject's experience. While looking at the initial image, the observer is primed in a number of ways and enters into certain structures of anticipation. I will explore how these anticipatory structures carryover from the initial image and impregnate the subsequent after-image. For what is an after-image if it is not the carryover of impressions formed during the initial sensory experience? A state of anticipation is a particular way of being, which I refer to as the presence of intent.

The carried-over experience that forms the after-image

To form an image of an object (whether it is a recognisable object or just as a blurry blob), the impulses received must be transformed into an understanding. To employ a metaphor, the package received must be opened to reveal its content. This is very different to the situation where the initial impulses that impinge on our visual system are not influenced by intent and nothing is required of them. No content is sought. They simply travel through the optical system undisturbed in a natural flow between the physical universe and our visual sense organs. No perceiving subject is involved; no one achieves an outcome. However, with the intent to produce an image, we intervene in the flux of impressions and structure them. Units to be compared and contrasted are formed, such as length, angles, placement, and brightness. Relationships are established that enable the subject to achieve consciousness of a visible object. Of all the structures that form and enable a subject to see a visible object, which of them are carried over and influence the after-image?

Not memory, but an ongoing presence

First, let me use two examples to clarify what is implied by structured attitudes that can carryover from the initial image to the after-image.

The first example illustrates how a visual experience carries over when the initial situation is no longer present. When we travel by train and look out of the window at the passing scene, that sense of movement remains when the train stops. The scene outside the window appears to slip backwards. This common experience is a well-known illusion that gives us the strange sensation that comes from knowing the train has stopped and yet the view keeps moving. The previously accumulated experience of movement while the train is going forward is prolonged. We impregnate what we see in the present with what we previously saw.

The second example shows how a similar prolongation is also present in auditory perception. As we listen to a piece of music that has tonal structure, we settle into a form of hearing with specific parameters. One of these is key structure. We sense the relationships that set the music in a certain key and if a note breaks with the given key structure, we notice it instantly—ouch! It is disturbing. We immediately recognise the intruder. This implies that we carry within us the key structure that determines the way the music sounds. It is this experience that I describe as 'carryover'. The presence that attunes us in a structured way to what we hear does not begin or end with the specific notes as they are played. It is an overarching presence. For instance, a symphony can modulate through different keys but the work ends in the key with which it began. Without this, we'd be left hanging in a state of unresolved suspense.

This prolongation that aspires to a sense of balance is not memory. It is not a process of recall. Rather, it is an ongoing presence and can be understood to include everything that animates and forms the immediate experience. This is the experience we have of *carrying-over*, through time, content that is present in our way of being alive at that moment. That which carries-over is out of reach of our conscious manipulation and yet it influences how we perceive what is to come.

Forms of anticipation that affect an after-image

Visual perception draws from more than the basic interactions described by the physics of light and the biology of the physical body. The process is necessarily subjective, in that the observer's attitude and anticipation form the carryover that affects what is seen. Anticipation can have a variety of forms, of what I call *structures of intent*. The following description of a number of these casts some light on how they produce carryover effects that generate the after-image.

1. The determination of brightness values

Contrast enables perception. Without contrast there is no perception. Many experiences bear witness to the dissolution of perception where contrast is lacking. In viewing an undifferentiated illuminated surface that fills the entire field of vision, the observer finds that, after a short time, all sense of colour is lost and even the sense of how bright it is. Other senses show a similar loss of discernment in an undifferentiated medium. Seeing an object implies that at least two brightness levels are present. Different areas confront one another and the contrasting values are integrated to establish a scale of brightness across the field of view. The realisation of interrelated ratios of brightness brings visibility. With this, looking results in seeing.

A well-known example is the Adelson checkboard made up of light and dark grey squares (see Figure 3). Two squares, marked A and B, are the same shade of grey when seen side-by-side, but when they are separated, one appears much brighter than the other. This happens because we do not see an absolute measure of the rectangle's luminosity but the result of the integration of contrasting brightness values in the areas surrounding the rectangles.

The same effect can be observed in coloured images. This is illustrated in a second example (see Figure 4). It shows that stripes that appear to be the same colour when joined together take on different colours when they are seen in different colour contexts. The reason for this is exactly the same as in the changed view of the two grey rectangles. The colour that appears is formed by the contrast in luminosities (light or dark grey) registered by each cone set – as described by the unified principle of colour.

The Outcome: This example illustrates the carryover effect. The ratios of luminosities established in the initial image are the basis for what we subsequently see, that is, for what we intend to see. When the initial image is removed and we see an after-image on a blank page, we're carrying forward the already established brightness ratios. These will stay with us for a short time, as long as the relative values of brightness formed for the previous image have not been effaced by a different view. We look at the blank page anticipating not only the same pattern, but also the same ratios of dark and light from the initial image.

2. Looking with the anticipation to see

When we look, we are animated by the expectancy to see something. Seeing something confirms the impulse that animates looking, but the expectation that drives looking does not end with the image.

The Outcome: When the initial image disappears, the intention to see continues within us and we look to see what appears next. This expectancy functions to call up an image when we look at the blank surface.

3. 2D surfaces

Forming a two-dimensional view is not simple and requires a number of parameters (see also Appendix II). It is the understanding of a 2D expanse that allows different areas of luminosity to be co-present side by side. We take co-presence and the spatial continuum for granted, as if what we see is merely reflecting what is 'out there'. But it is we, individually, who generate the understanding of surface and volume that enables us to receive the sensory data.

The Outcome: The continuum we recognise as an extension in space is an attitude that is present within us and that also becomes part of our anticipation when looking at the blank page. It does not drop away when the initial image is removed. We continue to anticipate seeing a 2D display. We are primed to accept any impressions that appear as belonging to a continuous extension that forms an integral display.

4. Objectivity

The projection of a two-dimensional extension generates a sense of objectivity. What appears is understood as 'other', as not being me. We recognise it as an autonomous, separate presence, whether produced solely in the mind or in the world.

The Outcome: The sense of separation from self is a profound attitude we cultivate when looking at what we understand to be a world 'out there'. This has to do with producing the intuition of objectivity.[50] Once present, this attitude does not disappear. When the initial image is removed, our attitude of relating to something out there of its own accord is prolonged and gives substance to the after-image.

5. Duration

We attribute a sense of duration to what is seen. Psychologically, we settle into seeing something and looking becomes a prolonged activity. Within us, we hold a sense of 'ongoingness', of temporal continuity.

The Outcome: When the object is removed, the 'momentum' of duration, of continued seeing, remains with us. The sense of a temporal succession unfolding carries over from before the object was removed, and we look at the blank page to see an object that has temporal presence and is there to look at.

In conclusion

Given such factors, it is no surprise that the after-image is not a passive response to physical and neurological processes in the body. The subject's understanding has a real effect on what manifests. The observer's acquired knowledge determines the formation of the after-image. Understanding carries over. The observer's previous achievements are present and active in structures of intent that are active in what is seen. For this reason, a description of the process that leads to seeing after-images necessitates the inclusion of subjective

---

[50] The acceptance of objectivity is taken so much for granted that realising it is a construct may come as a surprise. We might think we always knew the world was an objective fact but, in truth, it is a cultivated point of view. Appendix I offers a glimpse of what is overlooked.

factors. They have their source in being, in the intelligence and the sensitivities of an observer intending to achieve a certain end – in this case, sight.

I propose that these anticipatory factors constitute the effect that drives after-image production and that lies behind the reversal of bright and dark in the after-image.

The reversal

In seeing the initial image, certain retinal cells register light and others dark. To produce an image, the light must register as being lighter than the dark, and the dark as darker than the light. A relationship is created. But where does the understanding of this relation take form? It implies that while seeing what is white, I sense the contrary and, while seeing what is black, I sense the brightness from which it differentiates. Without this, I would not see a white object nor a black object. And yet, when I look at what is white in the image before me, I am not conscious of seeing anything at that place but white. I do not see or think of black. Where does the sense of contrast come from? On some level of expanded consciousness, there is a recognition of the white area contrasting with the black. I propose that when we see white, the knowledge of the contrasting black is present on the non-conscious level of the expanded mind, and vice versa when looking at black. The counterbalance to what we see consciously is absent to the conscious mind while it is present to the non-conscious mind. What we see consciously and what we sense non-consciously balance one another.

Support for this proposition can be found in the normative pulse rate of the optic nerve. Light and dark are not absolute measures of the light that is present or that is not present, but are in a mutual, relative relationship. In darkness, the pulse rate in the optic nerve increases, in light it decreases. The pulse rate is thus a baseline for comparing different effects. The presence of darkness just as presence of light is a relationship, a quality that can be more or less intense. Sight is not the result of objective factors presented to the eyes. Rather than simply registering the presence and absence of light, the perceiving subject brings together relationships that generate a meaningful comprehension.

In various ways, what we retain non-consciously, influences what we see. The mind has a great influence on what the eye sees, as the following simple experience shows. We wake from a night's sleep with our eyes still shut, but with the sense that it is morning. Through our closed eyelids we see the morning light that must be in the room. We open our eyes to find it is still night and all is dark. We close our eyes again, the light we imagined we saw is no longer present. Attitude and expectation influence what we see to a large degree, but this simple anecdote illustrates how anticipation can influence what we see.

A vast amount of non-conscious knowledge gathers in our expanded consciousness and enables us to recognise every item of which we become conscious. Conscious knowledge is the miniscule tip of the iceberg that is the activity of perception. Awareness deals with so much more than that which reveals itself to our conscious mind. For instance, seeing an expanse of water at a distance is a momentary conclusion we register consciously. But in so doing, we engage a list of factors so long that would fill this small book! The glistening area has to be recognised as a horizontal surface and not a strip of light; that it is at a distance also requires numerous clues, their interrelationships, and their integration. To know that water is what we consider it to be, requires a wealth of accumulated knowledge. I could go on, but the point has surely been made. We are oblivious to innumerable cues that never reach our consciousness and yet they sustain and give meaning to each momentary perception.

It should therefore be no surprise that, *in potentia*, we hold a huge amount of essential knowledge, available to integrate into every act of perception. This massive influx of impressions is our natural state of being beyond our linear, conceptual consciousness, in another kind of awareness, which I refer to as the expanded mind. It can be understood as the activity of multi-consciousness and includes an unrestricted number of factors working together to produce and uphold every moment of perception. This activity is generated by our purpose-bearing nature, which animates all towards certain outcomes. It is a process in which impulses that register upon our nervous system, but do not enter consciousness, are integrated, *multi-processed* one might call it, out of consciousness and become the objects perceived by the conceptual mind.

I argue that this activity is neither self-contained, nor self-driven. The activity of perception is closely bound up with the individual subject, and is animated by the intent that lives within that individual being.[51] And then, quite distinct from intent, are the forms of mind activity. Our experience of phenomena reveals that our mind functions in two different modes. One we know as conscious activity, which I refer to as conceptual consciousness; the other is the subliminal activity that bypasses conceptual consciousness and we only recognise its activity through inference.[52] Both modes affect our being and our understanding, but they function differently. Lack of a clarity in the distinction between these two separate mind processes underlies a tangle of unnecessary arguments on the subject.

---

[51] By *intent,* I refer here to what I have described (elsewhere) as a presence that is independent of thinking. Although with thought, we can sometimes recognise our intentions and name them, they form on a different and ineffable level of our being that influences the dual mind processes I describe in this paragraph.

[52] These two modes are defined in Chapter 2.

Now, having brought attention to the concert of the non-conscious and conscious mind in every act of perception, it is possible to return to the discussion of after-images.

The reflex action of after-images

The tendency to prolong structures formed when viewing the initial image is understandable in the context of carry-over effects. The anticipation to see something is present and it carries different qualities. To form and to see a visible object requires contrast, as has been noted. Brightness contrasts that were installed while viewing the initial image reemerge, only now with the surprising reversal of light and dark.

Imagine a situation in which someone throws you a heavy object and spontaneously, in order to catch it, you tighten your muscles and ready your arms to receive the weight. But you are taken by surprise and the object turns out to be really light. You discover this as you catch it and your arms involuntarily spring up and might even bounce the object back into the air. This reflex action is the result of the mistaken anticipation.

After-images can be considered a reflex action. With the removal of the initial image, attitudes that were generated have not been interrupted and are present within us, but there is no visual presence to absorb the effects. On the one hand there is a blank page to look at, and on the other there is our anticipation of the structures that permitted vision to function and produce the initial image. We look outwards, anticipating sight of some two-dimensional display 'out there', with similar contrasting luminosities. These expectations continue to influence our vision and reverberate through our looking – in a similar way to the key signature of a musical composition influencing the notes we hear. This is an example of our active involvement as a perceiver. I have argued that the observer is not a mere witness to an independent reality pageant. It only appears so because consciousness is oblivious of the process that forms its content. Beyond the conscious activity of the mind, the duo 'brighter than' and 'darker than' are co-present. Only afterwards, in consciousness, do we separate them into side-by-side features, each appearing as if autonomous. But the brightness of all we see is known by its opposite, the white is understood as 'not black', the black understood as 'not white'. The same must be true of the intermediary brightness values, where dark grey is contrasted with light grey. With our conscious mind we see dark and light forms spread across a two-dimensional expanse. However, in registering what we see, the wider expanded mind, functioning on a non-conscious level, recognises each area of brightness in relation to its opposite. The object we perceive is always a meeting of opposites, between that which enters the conscious conceptual mind and the active knowledge held by the multi-conscious, subliminal mind.

The presence of opposite luminosities is active in every image we form. Seeing encompasses an undisclosed tension, undisclosed because the complementary part of what we see does not enter consciousness and is retained in the expanded mind. The brightness we see manifests because of its opposite beyond the threshold of consciousness. The white area we observe is able to enter consciousness by retaining the black counter-stimulus against which it is measured; the black area is recognised through contrast with the white counter-stimulus. There is a tension of opposites. Think of holding a cork under water. The moment we let it go, it shoots up to the surface or even bounces out of the water. Such is the response when we liberate seeing from the tension of opposites that is present in image formation.

After looking at the initial image and then switching to look at a blank page on which there is no differentiation, we continue to sense the contrast in the initial image between light and dark, but we are unable to attribute these brightness distinctions to the blank page. Its effect continues to be active. The tension that created the initial image has no object to which it can be attributed. With the removal of the initial image that sustained the separation of light from dark and dark from light, this tension is set free. Rather than simply fade away, the abrupt release propels the withheld, complementary, submerged view into our conscious mind as a visible object.

It may well be that this tension between what is seen (in conceptual consciousness) and its implicit opposite (in expanded consciousness) builds gradually as we look at the initial image. Holding our vision fixedly to one point on the initial image allows the tension between opposites to rise steadily and, for this reason, after-images do not form when we just look momentarily at the initial image. In order to produce an aftereffect, we need to look for some time. Apparently, eye movements can release the buildup of tension and therefore, to avoid dissipating the tension, we need to look steadily at one point on the initial image.

Up to here

In this way, I propose that the phenomenon of reversal is not due to physiological reactions but to a state of mind anchored in the subject's anticipation. I hope this description will open up a new avenue of thought, even though it is still far from worked out in detail. A process of rebound seems to fit better with our experience of after-images. It also suggests a path that breaks away from explanations based entirely on physical systems. Vision, like all sensory activity, involves far more than causal processes in the physical body. It includes the subjective involvement of an intelligent perceiver, an individual, living subject engaged in producing the observation.

$$\text{PART II}$$

The Benham Top: From Black and White to Colour

This part is comparatively long because it is a detailed account of a complex application of the colour-forming properties already described. It is included because the appearance of colour on a black and white spinning top offers a propitious opportunity to bring together what has already been presented. Armed with the unified colour principle, we are able to meet the challenge of understanding the appearance of colour under different circumstances. This part consists of two chapters:

Chapter 15 describes the Benham top, considers the tendency to search for explanations in false directions, and includes a section on the reason why there are significant variations between individual viewers.

Chapter 16 describes the observations and deduces how they are produced.

If the reader wishes to see the spinning Benham top in action, there is an example at: https://michaelbach.de/ot/col-Benham/index.html

15
Initial Observations
on the Spinning Top

ABSTRACT 15

- The Benham top has black and white markings on it. When spinning, it shows colours.

- Colours form on the Benham top, despite the fact that full-spectrum light reaches our eyes (see Chapter 1).

- In every situation, the colours we see are a genuine manifestation of colour. Each is equally objective, equally subjective. Nothing justifies separating colours into more or less real.

- Divergence in the colours seen by different observers is attributable to the fleeting nature of the colour-producing moment on the Benham top. It accentuates differences in an individual's rate of image formation.

- Individual differences are also the result of differing sensitivity to peripheral effects.

- Spatial (or visual) periphery is a recognised phenomenon, while temporal (or auditory) periphery receives little consideration. The latter can significantly influence the appearance of colour on the Benham top.

- Despite the variation in colours seen by different observers, agreement is more easily found when colours are divided into the blue half-spectrum, the red half-spectrum and those with a greenish tinge.

- The following observed effects tend to lead the search for answers on a false path:
  1. *Trap A* - considering the flicker pattern to be a colour code programme in our brain.
  2. *Trap B* - attributing colour to wavelength, as if colours issue from specific wavelengths of light.
  3. *Trap C* - separating the appearance of colour into less real (subjective colour) and more real (objective colour).
- Chapter 15 covers the preliminaries; Chapter 16 looks at how the arcs become coloured circles.

15.1 The Benham top

Late in the 19th century, a toymaker in England named Charles Benham produced spinning tops with a black and white design on the flat upper surface. When these designs are spun, they change from black and white to colour. Benham published articles on this top, including one in *Nature* in 1894[53]. The top and others like it are now known as Benham tops. Since that time, very many articles have appeared in journals and there are even books on the curious phenomena of the Benham top and yet, it still hides a secret for the transition from black and white to colour has not been fully elucidated.

The flat upper surface of the Benham top is divided so that one half is black and the other half is white. When it is spun, the two halves create a flicker effect. There are various models of Benham's top with different markings. In many of them, the white half of the disc has a number of slender black arcs at different radial distances from the centre.

Figure 9 shows a typical design, with four groups of arc-segments across the white half-disc: two groups have one end touching the black

---

[53] C. E. Benham, 'The artificial spectrum top,' *Nature* 51 (1894): 113-114.

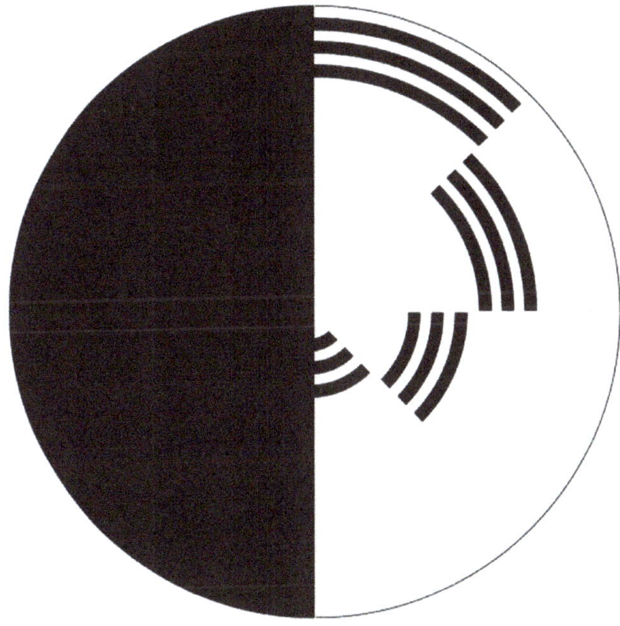

Figure 12. A common form of the Benham top

half; two groups are separated from the black half-disc by a white section. The arcs are one-eighth of a circle in length, in a staggered sequence on the white half-disc. When the top spins, these arcs appear as differently coloured circles.

As no definitive explanation of colour formation on the disc has been accepted, it is intriguing to view this puzzle in the context of what has been presented in this book about colour vision.

15.2 Thinking it through

To draw attention to the surprising part of what happens, here's a thought experiment. Let us momentarily ignore the fact that we already know what happens when the top is spinning. Instead, before we set it spinning, try to imagine what will happen. First, consider a top without the arcs, where the upper surface is simply half white and half black. What will this disc look like when spun? We expect the black and white halves to blend into a uniform grey. This does happen when the top is spun rapidly, somewhat faster than we can manage by hand.

Now, if we add the black arc-segments to the white half-disc, what should we expect? Surely, we expect them to form circles that are a slightly darker grey than the background produced by the half-black/half-white disc because, at the arcs' radius from the centre, they add more black to the background. The circle is black for five-eighths (62.5 per cent) of the time, and a little more than the 50 per cent of the half-discs without the arcs.

But this is not at all what happens. When we spin the top, surprise! The arcs appear as coloured circles.

15.3 The initial conditions

There is a fact that differentiates the Benham top from most other colour-producing situations. In the other cases discussed so far, we experience colour where the eye receives light from selected wavelengths of the visible spectrum. However, the Benham top is illuminated with full-spectrum light.[54] When the disc is not turning, we see the pattern on the disc in black and white (see 3.5) because the full-spectrum light reaching our eyes from the disc does not engender conflict between the three sets of cone cells. But on the spinning top we see colour, and yet wavelength differences are not involved. This raises the question: how does spinning the top cause conflicting brightness levels to register on the three cone sets?

---

[54] It is possible to shine a light that is not full-spectrum on the Benham top and observe colour effects, but these are variants that fall within the general principle.

15.4 A reminder: colour is not wavelength

The description of the process of colour formation throughout this work does not attribute colour to the wavelength of light. In case the reader is still haunted by echoes of this misattribution, I restate the matter.

All cases of colour come from light and dark encountering the creative perceptive abilities of a perceiver, of you, of me, and of every being that sees colour. Colour vision does not exist until the individual subject makes it manifest. Wavelength differences are not the source of colour. They play a role because they interact differently with the eye's light-sensitive cells but this serves solely to produce colourless intensity differences of darker and brighter between the cone sets. My argument throughout is that colour only enters a scene at the moment the being that perceives the colour creates it. It is an achievement of the mind in the same way that a sense of depth is an achievement of the mind – and, yes, this implies that mind is a capacity that accompanies every living organism.

15.5 Equally subjective, equally objective

I repeat the claim made throughout this work, that the process which elicits colour is the same in all the familiar cases where colour is seen. Dividing the phenomena of colour into different categories is not helpful, although widespread. For example, the reference to 'subjective colours' in neurobiology and elsewhere, or 'physiological colour' by Goethe. The implication is that colour is sometimes more objectively real and at other times less so, as if there were counterfeit colours that mimic real colours. Such divisions render the colour phenomenon unnecessarily complex and are misleading in attempts to understand the Benham top. Colour is what we see, whether it is a green leaf, a rainbow, a coloured shadow, or an after-image that appears in our mind after staring at a coloured surface. In all these cases, the colours we see are genuine and are equally subjective, equally objective.

In holding to the viewpoint that colour results when the cone sets register conflicting brightness levels, what we need to question with the Benham top is: how does a black and white, spinning top produce contradictory views between the three cone sets? The research objective is as simple as that!

15.6 The spin is not a factor in colour production

It is noteworthy that the movement of the spinning top does not influence the colours we see. The spinning itself is not what generates colour. Colour results only from the alternation between light and dark. This is known from an experiment which produced the flicker sequences of the Benham top on a static image without the spinning movement, instead using projectors that were switched on and off rapidly.[55] In that experiment, the colours that resulted were the same as those that arise with the spinning top, which demonstrated that sequenced patterns of dark and light alone produce the appearance of colour.[56]

15.7 A description of the Benham top when spinning

This refers to the top pictured in figure 15.1, with four groups of arc-segments on the white half-disc.

---

[55] L. Festinger, M. R. Allyn, C. W. White, 'The Perception of Color with Achromatic Stimulation,' *Vision Research* 11, no. 6, (June 1971): 591-612.

[56] There is a factor that I have not included. According to the same article, the tiny moment of transition from light to dark and from dark to light is not an instantaneous switch. The experiment showed that the clearest colours resulted when the almost instantaneous change of brightness followed a certain gradation. It is possible that this graded transition could be a factor in the conflict that resolves into colour, which is certainly of interest. However, since the process I describe answers the central query, whatever further influence there might be would not be a determining factor.

1. Take the case of an arc that begins where the black half-circle ends. This arc-segment extends the black of the half-circle for another eighth of the turn. After this, the circle at that radius turns white for another three-eighths turn.
2. At that arc-segment, there is a change **from black to white.**
3. Now take the arc-segment beside that one, that begins after one-eighth of a turn of white. This means, that when the black half-disc ends, there is one-eighth of a turn of white, then one-eighth with the black arc, and then two-eighths of white.
4. The change at the arc-segment is **from white to black and then from black to white**.
5. The next arc-segment begins two-eighths of a turn after the black half-circle and there is one-eighth of a turn of white after it before entering the black half-circle.
6. Here, the change is also **from white to black and then black to white**.
7. The last arc-segment starts after three-eighths of the white half-disc and its one-eighth turn ends with the black half-circle. The change is just **from white to black**.

These changes are significant, as will be shown in Chapter 16, in that the passage from black to white yields hues from the red half-spectrum; the passage from white to black yields hues from the blue half-spectrum.

The four states described above are the changing white-black sequences at the arc-segment as the top spins in one direction. In these four positions the largest agreement between observers describes: 1 = reddish, 2 = yellowish, 3 = greenish-blue, 4 = darker blue. When the top is spun in the opposite direction, the sequence of colours that appear is reversed: 1 is now darker blue to 4, which is reddish. Why does the colour sequence reverse when the spin is reversed? And what is it that restrains white and black from blending into an intermediary grey? These questions will be addressed in Chapter 16.

As to the function of the black half-disc, that will come in later as it does not directly influence the colours that appear.

15.8 The colour of the arc-segment

A typical view of the Benham top shows arc-segments in a staggered sequence on the white half-disc. However, when several arc groups are removed and only one arc-segment is left on the white half-disc, that single arc forms a circle that is the same colour as was produced by the group of arcs in that position. The colour produced is not dependent on the sequence of staggered arcs even though, when seen together, the effect is more impressive.

15.9 Variations between observers

There is significant deviation between observers identifying the colours that appear on the Benham top. Different people see different colours, and at times only hesitatingly and at other times quite clearly. The practice time they spent viewing the visual impressions generated by the Benham top and even their mood and state of being seem to have an effect on the visibility of colours. However, we should not conclude that the observations lack all structure. Within the many variations, there are fundamental colour distinctions upon which many observers agree. The exact colour hue observed may vary, but the variations are greatly reduced when one considers the fundamental distinction between the two half-spectra (see 5.2). There is a reasonable consensus when classifying the observed colours as belonging to the red or blue half-spectra, or to their combination as a greenish colour.

Of course, the question remains: why does the observed colour vary so greatly with the Benham top in comparison with other situations where colour is observed? The reason for the discrepancies between observers must be due to a large extent to individual differences in the rate of forming an image. The colour seen is a momentary production. Any minute difference between observers in the rapidity of transduction of the visual impulse and in the image formation would have a major effect on the colour and on whether a colour is seen at all. The momentary nature of colour formation on the Benham top is described in Chapter 16. In other colour vision situations, the conflict the colour phenomenon resolves is due to light that contains only part of the full spectrum. The imbalance persists over time. Different observers can resolve the visual effects into colour at their individual rates, which ensures a greater agreement. The fleeting moment of colour formation on the Benham top leaves a much greater scope for variation between individual observers.

15.10 Observer variation due to peripheral effect

A further cause for variation concerns periphery perception. This is more technical and I only mention it here briefly as variation in periphery effects leads to a whole new subject that is more fully discussed in *Making Sense*.[57] When we look at something, we bring it to the centre of our visual field. This fovea view is embedded within a wider view, the spatial periphery that brings other qualities to what is seen. The effects of fovea and periphery function together to produce all we see, yet sensitivity to peripheral phenomena varies between individuals.

In the same manner that there is visual periphery (spatial periphery), there is auditory periphery (temporal periphery), although its presence is largely unrecognised or misappropriated. It is not formed by spatial displacement but by temporal displacement. That

---

[57] *Making Sense*, Chapter 2, 'Sense Perception: Sight and Hearing', in particular, pp. 56-64 (see footnote 3, p. 4).

which precedes and follows the sound that reaches our ears constitutes the periphery. As in the case of visual periphery, in auditory (temporal) periphery, all is co-present.[58] The wake left by what was heard and the anticipation of what is to come, are co-present with the sound that we identify as happening in the 'now'. For instance, the words of a phrase impact together on our understanding. The sense that we recognise is, in most cases, not built sequentially word by word.

Through temporal periphery, the momentary view of what we see always contains more than what registers on our visual sense organs at that moment. This fact is obvious since, if seeing did not include more than what constitutes the actual passage of what we call the present moment, it would have no extension. It would be nothing but a mathematical point of zero dimension or a line with no width and, in that case, nothing would be seen! Co-presence in a temporal dilation generates all we perceive.

Temporal periphery perception varies dramatically between different observers and therefore what is perceived as co-present must also vary. Such differences are not significant in most situations. Colour usually resolves a continuing state of contradiction that does not change from moment to moment. In the case of the Benham top, however, the fleeting passage of the arc segments produces exceedingly short-lived contradictions. Differences in peripheral sensitivity have a major influence. The time it takes for the conscious mind to achieve perception will influence what is seen. For example, a previous impression (seeing white or seeing black) can sustain a different level of presence for different individuals. If the observer strongly retains the presence of temporal periphery, its impact on the moment of integration that produces what is seen will be different. Differing sensitivities

---

[58] As demonstrated in *Making Sense*, this expansion of perception beyond the immediate moment does not involve memory.

to peripheral presence will produce different luminosity contradictions, or even the absence of contradiction. This leads to the conclusion that the conflicting views that arise in the mind of an observer differ from person to person and necessitate different colour hues to achieve a resolution. In some cases, temporal differences go as far as reversing the luminosity relationships between the cone sets.

Given the great range of possible variations between observers, the degree to which there is agreement about the colour we see on the Benham top is surprising. Apparently, we are not so terribly different from one another – as suggested by the broad agreement among a significant number of subjects about which half-spectrum is being observed.

15.11 Traps that ensnare attempts to understand

When taking account of the factors that have a role in the production of colour on the Benham top, it is very easy to take a mistaken path and arrive at a false conclusion. There is the flickering background produced by the black and white halves of the disc, and the particular flicker pattern of the black arcs that become coloured circles. These flicker patterns are determined by the position of the black arcs on the white half-disc and each pattern corresponds to the appearance of a certain colour. There is a temptation to draw a causal relationship between the light-dark flicker sequence and the appearance of colour; to interpret the sequence as a code, almost a Morse code, that automatically triggers an embedded programme in 'the brain'. The brain is then confused (fused together) with the living subject that perceives a particular colour.

The conclusion that, in colour production, the brain functions like a computer with a software programme, leads astray the search for understanding. I suppose this trap is set by the pervasive belief that sensory responses are produced by programmes installed in our neural network. To go one step further, I wonder if this interpretation is not retained as a safeguard for an undeclared materialist view that attributes mind activity exclusively to physical events. To set such a view to the side, even temporarily, can prove helpful when considering how colours appear on a spinning Benham top.

15.12 A caveat

At this stage I propose a hypothesis-on-trial because I do not have a full range of supporting evidence built on experimental work with the spinning top. Even without this, given the observations described in previous chapters, I believe the proposed approach leads to a straightforward answer to how the spinning Benham top becomes coloured. It proposes a unified description of colour formation, fully coherent with other experiences of colour.

With these clarifications, a fresh path to an answer lies before us, and it is taken up in Chapter 16.

16
The Process of Colour Formation on the Benham Top

ABSTRACT 16

- Image formation (Chapter 4) is accomplished only where a brightness relationship is established between dark and light. This happens where the black arcs are bordered by white.

- When there are no black arcs, the spinning black and white top appears grey. Black arcs are crucial to the appearance of colour.

- The position of the arc-segment determines the timing of the dark/ light transitions.

- This fact tempted researchers to interpret the flicker effect of the black arcs as a code on the assumption that our visual sensory system is preprogrammed to decode these flashes and transform them into colour.

- This assumption is erroneous, according to my understanding.

- Colour is generated on the Benham top, as elsewhere, to resolve cone views with conflicting brightness.

- Some research finds that the rate of transduction of visual information is more rapid for longer wavelengths than for shorter wavelengths (see Chapter 8). From this we can expect that the cone sets form their views at different rates: the L cones most rapidly, then the M cones, then the S cones.

- Colour is required to resolve concurrent contradictory brightness values and a flash of colour, repeated through the circle at the radius of the arc-segment, forms a circle of colour.
- The general principle:
  1. Passage from the black arc-segment to the white disc yields hues from the red half-spectrum.
  2. Passage from the white disc to the black arc-segment yields hues from the blue half-spectrum.
- The variation of the hues in each half-spectrum from lighter to darker depends on the position of the arc relative to the black half-disc.
- The colour tends towards green when the entry into and exit from the arc are balanced. This happens when the arc is placed at the centre of the white half-disc.
- The black half-disc is not directly involved in colour production, in fact, a white half-disc in its place shows similar colour circles.
- The black half-disc appears to function as a pause in the sequence, separating the individual, colour-forming cycles.

16.1 Prerequisite information

1. (Chapters 2 and 11) Even though the conscious person observing the spinning disc is oblivious of the rapidly passing forms on the top's surface, these forms are precisely recorded by the expanded mind.

2. (Chapter 4) Perception is a momentary act, not an ongoing process. It requires the repeated, momentary formation of an image which integrates all the visual properties in the field of view, such as position, dimension and brightness,

3. (Chapter 1) Image formation is realised separately by each cone set and the three views are later fused to become what is seen.

4. (Chapter 3) If no conflict in brightness levels between the three cone sets is present, the image is colourless, that is, in shades of grey (black and white).

5. (Chapter 3) When the brightness levels recorded by the three cone sets contradict one another, the conflict must be resolved in order to produce a visible image.

6. (Chapter 3) Colour has a property that permits differing brightness levels in the different cone cells to form a single view.

7. (Chapters 3 and 12 and Appendix I) This single view is achieved by moving beyond a mentality that perceives levels of brightness, to a new way of seeing that recognises concurrent ratios of brightness

8. (Various chapters) This new relationship with what is looked at is achieved through an imperceptible shift in consciousness that transitions into colour vision.

9. (Chapter 5) The spread of prismatic colours comes from two separate half-spectra, the violet–blue series and the red–yellow series, that are formed separately and contain complementary colours.

10. (Chapter 9) Research into the transduction of light passing through the retinal cells and to the optic nerves suggests that longer wavelengths are processed more rapidly than shorter wavelengths. As each cone set is centred on different wavelengths, this implies that they accomplish the task of image formation at different rates.

11. (Chapter 9) Given the differing rates of image formation between the three cone sets, the L cones would be the first to form an image, followed by the M cones and then by the S cones.

12. In order to understand colour formation on the Benham top it is essential to recognise the reasons why different people's experience varies greatly, as discussed and set out in 15.9 and 15.10.

16.2 The approach

The pattern on the Benham top is black and white and yet, when the top spins, what was colourless appears coloured. Since I argue that colour results from the clash of different brightness levels, the question is, what produces the contesting views in the three cone sets, thereby generating the need for a resolution through colour vision? I believe the answer lies in the fact that different cone sets integrate visual information at different rates (Chapter 9).

16.3 Arc segments and the circles they form

When the top is spinning, the view of the arc lasts only a fraction of a second. If the arc is one-eighth of a circle and the top turns ten times a second, the view lasts one-eightieth of a second before it changes. At these, and even at slower, rates of change, the time variation between the three cone sets processing an image will have an effect (see point 10 in 16.1). The formation of the three images will be out of step. At the moment in which the three views fuse (integrate) to form a conscious image, they will not be registering the same scene. See 16.9 for a full description of how these differences create contradictions in brightness that yield colour.

To state the obvious: when the top is *not* spinning, we see the markings as black and white. Even though the L cones are integrating the image at a faster rate than the S cones, the time difference has no significant effect because the view of the static top is not changing over time and thus the minute differences in the rate of image formation do not influence what is seen.

When the top is spinning, what is only an arc segment appears as a full circle. This does not surprise us because we are familiar with the phenomenon, for example, a light on a turning wheel traces out a circle or we can draw circles in the air with a sparkler. Even though the expanded mind can register the sequence in great detail (see examples in Chapter 11), the rapid sequence of stimuli surpasses what the conscious mind can follow. In the case of the Benham top, the moment of integration and entry into consciousness is a flash of colour. It is produced sequentially at every position along the circle, that is to say, it is picked up sequentially on a circle of different retinal cells. In this way, the coloured flash registers in rapid succession at every point along the circuit, delineating a coloured circle.

16.4 The black half-disc as a reset function

The role of the arc-segment will be clarified further, but what are the roles of the black and white half-discs? Image formation requires that the three cone set views are integrated at a particular moment.[59] This takes place during the passage of the arc-segments on the white half-disc. During the passage of the black half-disc, visual activity subsides. Thus, there is a pause followed by a new cycle of impressions from the arc-segments. This suggests that the passage of the black half-disc is not directly involved in generating colour. The same conclusion is suggested by the fact that the half-disc without arcs does not need to be black. It can be white (but not exactly the same shade as the white half-disc) and the top, when spinning, will produce the same or similar colours. We can surmise that the half-disc without arcs functions as a pause in the image-forming process and is not a requirement for colour production.

16.5 Image formation at the arc-segment

On a spinning top with no arc-segments and only a white and a black half-disc, colour does not appear. When one or more black arc-segments are added to the white half-disc, colour does appear. What

---

[59] Chapter 11 describes how, when we view a film (a movie) and see a continuous moving record of events, there is an accumulation of momentary perceptions. In observing movement around us, the same phenomenon takes place but at many more images per second.

is the factor that produces colour when the arc-segments are present? This prompts the obverse question: why does no colour appear when the arcs are absent, that is, when there is only the black and white flicker of the turning half-discs? The short answer is: colour is absent because image formation is lacking. This relates to the process of image formation detailed in Chapter 4.

Image formation and the appearance of colour are concurrent and require the presence of luminosity relationships. The arc-segment enables the establishment of these brightness ratios between the arc and its immediate surrounding.[60] This establishes both the act of seeing an object, and the presence of an observing subject conscious of perceiving something.

16.6 Registering stimuli versus seeing an image

A clear distinction must be made between receiving stimuli that produce a visible image, and the reception of a flux of impulses without image formation. There can be optic stimulation that does not produce an image and thus does not involve a perceiving subject. In the case of the Benham top, our visual system registers the changing brightness of the two half-discs. When we look at a bright surface, the neural impulse rate responds in one direction and when we look at a dark surface it responds in another direction. However, this does not imply image formation. A decrease or increase in the neural response can be received passively, simply as a motor response that does not require relating the different states to one another and interrelating them into a unified view. Ratios of brightness do not form even though the changes influence the normative impulse rate of the optic neurons.

---

[60] The act of integration couples image formation with colour production. The fact that the ratio of luminosities produces colour became apparent in the description of coloured shadows (Chapter 6).

Because the compulsion to see an image is lacking, no conflict that demands a resolution is generated and colour is not called for.

The presence of the arc-segment creates a contrast between the dark line and its bright surround. This contrast stimulates the subject's expectation of a visual object. Image formation and the achievement of colour go hand in hand. The subject's anticipation of a visible object calls for the resolution that produces colour (Chapter 3) and the would-be perceiver enters as a conscious subject.[61]

16.7 Rate of spin

When the top spins faster or slower, the coloured circles are essentially unchanged. We might expect the rotation rate to shorten or lengthen the presence of luminosity contradictions in the views formed by different cone sets and therefore to influence the colour.

Why do the hues remain so nearly constant? The answer is that the determining factor is the momentary transition between light and dark at the ends of the arc-segment. With every turn of the top, whether faster or slower, the same transitions are enacted. This will become clearer in the following sections.

Above a certain rate, the colours fade away. This surely happens when a rapid passage of the arc-segment prevents the establishment of light and dark ratios. Without this, no integration of contradictions is called for and no image forms. The colour formation that takes place at lower spin rates, ceases.

16.8 An installed programme or a living being

I pause here to add a caveat. Those who believe in a 'robot-like' reaction of the sensory system might suggest another explanation for

---

[61] The fluctuating, non-continuous, presence and absence of the subject is considered in greater depth throughout *Making Sense,* pp. 203-4, in 'Multiple I's' (see footnote 3, p. 4).

why the same colour persists at differing rotation rates. They can claim that the pulse sequence, or code, remains the same when the rate of spin changes because it is simply played out faster or slower. I reject the assumption underlying this suggestion. It stems from the viewpoint that a programme engineered into the brain recognises the pulses as coded sequences. The supposition is that this code becomes a message that is relayed to a centre (or centres) within the perceiving subject, where it activates what is necessary for a specific colour, correlated with the coded sequence, to enter the perceiver's conscious mind and subjectively produce sight of that colour. This, in my own words, seems to be the premise. It is a view that relegates conscious perception to a functional link in a message chain, which is why I refer to it as a 'robotlike' response. The answer I seek implies a fundamentally different point of view, more fully described in the Epilogue.

16.9 Colour formation on the Benham top

I remind the reader that the start of Chapter 15 references a video on the web (https://michaelbach.de/ot/col-Benham/index.html) of a spinning top similar to the one described here.

The general principle

This first description gives the general idea that will then be considered in more detail.

In the transition from **the white background to a black arc-segment**, the S cone activity dominates because the slower responding S cones still register light, while the faster responding L cones register the image of the dark arc in relation to the surround.

In S cone dominance, we are able to resolve the contradictions in the three cone sets with a colour from **the blue half-spectrum**.

In the transition from **a black arc-segment to the white part of the disc,** the L cone activity dominates because the faster L cones pick up the change from dark to light, while the slower S cones still register the image of the dark arc-segment in relation to the surround. In L cone dominance, we are able to resolve the contradictions in the three cone sets with a colour from **the red half-spectrum**.

This description of how colour forms on the Benham top shows why turning the top in the opposite direction causes the complementary half-spectrum to appear. When the direction is reversed, the progression from light to dark and dark to light is reversed. Therefore, colours from the blue half-spectrum become colours from the red half-spectrum and vice versa. Reversing the direction changes a hue to its complementary colour.

The principle governing transformation into colour

ENTRY into the black arc generates the BLUE *half-spectrum.*

EXIT from the black arc generates the RED *half-spectrum.*

These two statements are the full story of colour formation on the Benham top.

What follows is some background information needed to describe the process in different situations. RED refers to the range of hues in the red half-spectrum, BLUE refers to those in the blue half-spectrum. For clarity and simplicity, only the S and L cones are described, as they demonstrate fully the principle involved. Adding the M cones would make the description more complex but would add nothing further about the process.

A step-by-step description

A) Arc-segments contiguous with the black half-disc

ENTRY into the black arc generates BLUE. EXIT from the black arc generates RED.

This is clearly demonstrated when one end of the arc is touching the black half-disc and only the other end of the arc is free to generate colour.

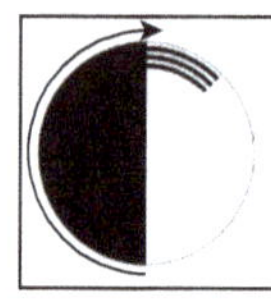 Case 1: There is a transition from white into the black arc-segment and no exit, because the arc-segment continues into the black half-disc. With only the ENTRY end of the arc that functions, we see BLUE.

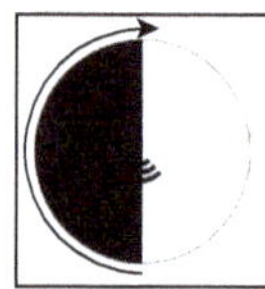 Case 2: There is no entry into the black arc-segment because the arc continues directly from the black half-disc. With only the EXIT from the arc into white, we see RED.

B) Arc-segments in intermediary positions

In intermediary positions on the white half-disc, the arc-segment is not joined to the black half-disc but is separated by a white area.

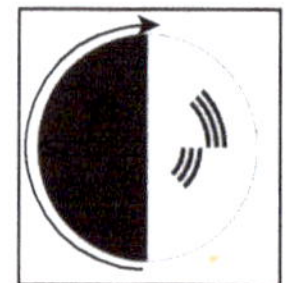 A freestanding arc-segment has both ends free to produce a transition between the white disc and the black arc. Both ENTRANCE into and EXIT from the arc produce a moment of colour that results in a blending of the two effects. The two half-spectra are brought together, one counterbalancing the other.

Since entry into the arc produces BLUE and exit from the arc produces RED, the result is a blend of the two.[62] To comprehend this, we need to take into account which effect is more intense. The more intense of the two will determine the half-spectra that rules, but the colour that appears will be partially neutralised by the presence of the weaker half-spectrum.

What determines which half-spectrum is more intense? At this point in my work, I propose no answer (see 16.10) and offer instead a detailed description of what we observe.

C) Quadrants

It is helpful to speak of the white half-disc as divided down the middle into two quadrants.

---

[62] A reminder: light has no colour. This reference to colour is to a set of brightness ratios (see 10.2).

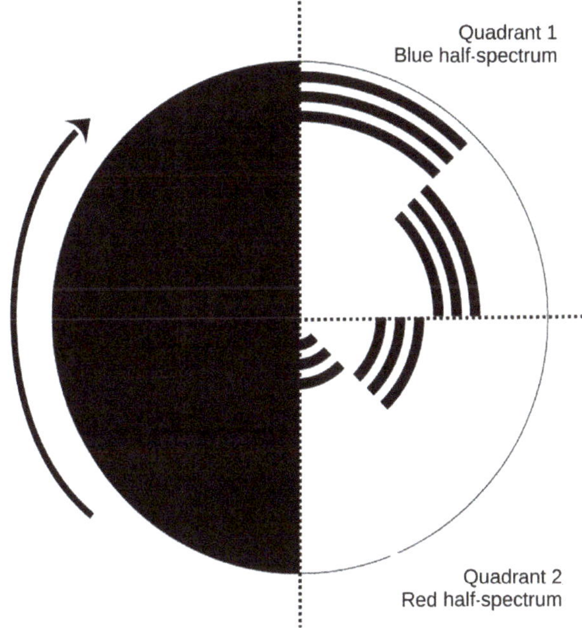

Figure 13. The quadrants

On a top spinning in a clockwise direction:

Arc-segments in quadrant 1, appear BLUE. Thus, ENTRY into the arc is more intense than exit.

Arc-segments in quadrant 2, appear RED. Thus, EXIT from the arc is more intense than entry.

D) Midline reversal

It is apparent that the midline between the two quadrants (in the middle of the white half-disc) is where the colours of one half-spectrum flip over to become those of the complementary half-spectrum. An arc-segment placed beyond the midline produces colours from the other half-spectrum. On one side of the midline, the entry into the arc is more intense than the exit; on the other side of the midline this is reversed and the exit from the arc is more intense.

When the arc is placed over the midline of the white half-disc and extends into the two quadrants, it takes on a greenish hue.

E) How position relates to colour

There is a simple logic to determining the colour that will appear when the top is spinning.

In quadrant 1: the stronger BLUE half-spectrum encounters a less intense red half-spectrum. As the arc segment moves in the direction of the dividing line between the quadrants, the blue colour will change, moving along the colour wheel away from violet towards light blue. The blue is diluted with brightness values that produce red, but that are not intense enough to displace the dominance of the blue half-spectrum.

In quadrant 2: the stronger RED half-spectrum encounters a less intense blue half-spectrum. As the arc segment moves towards the dividing line between the quadrants, the red colour shifts along the colour wheel, away from dark red towards yellow. The red is diluted with the brightness values that produce blue, without transiting into the blue half-spectrum.

To summarise: as the distance of the arc-segment from the black half-disc increases, the observed colours move either in the direction violet-blue-light blue or in the direction red-orange-yellow. The sequence is determined by the quadrant they occupy.

F) The colour circle in more detail

The progression of colours along the colour circle is related to the process of colour formation described throughout this book, namely, that different cone sets register incompatible brightness values.

Let us consider quadrant 2. The delayed response time of the S cones relative to the L cones (Chapter 9) creates the visual conflict at the arc ends. This gives rise to a colour from the red half-spectrum at the exit from the arc segment, and a colour from the blue half-spectrum at the entry into the arc-segment (see 16.9). The combination of different luminosities producing what appears as red has to incorporate the brightness values producing blue. That which at the exit of the arc appears bright to the L cones but dark or even non-existent to the S cones, changes in the encounter with the other end (blue-producing) of that arc-segment. There is an added quota of light absorbed by the S cones, which changes the mix of brightness values and counteracts the stronger response by the L cones by introducing more light to the S cones. The ratios of brightness change. The red effect is diluted. To resolve the conflict in brightness values and produce a single view, the colour must move from red towards yellow.

What happens in quadrant 1 is parallel and opposite to what happens in quadrant 2. The colour produced at the ENTRY into the arc-segment (blue) is diluted by the colour produced at the EXIT from the arc-segment (red).

In this way, it is possible to trace the shift along the colour circle and to describe how it results from the position of the arc ends on the white half-disc.

G) Light to dark, dark to light sequences in the cone triplet

Chapter 10 presented the light to dark and dark to light sequences in the cone triplet, when blue forms with ascending brightness values, and red with descending brightness values. It seems the same process functions in the case of the Benham top, even though the context is very different.

The blue half-spectrum forms at the entry to the black arc, where the L and M cones register black while the S cones register light (see 10.3). Thus, L and M cones respond more rapidly, ahead of the S cones, and a progression of ascending brightness forms in the triplet. In the case of the red half-spectrum, there is the opposite sequence: the brighter L and M cones exiting the arc-segment register light, while the last to form, the S cones, register the black arc. The sequence is one of descending brightness. The sequence from light to dark (where red is seen) and the sequence from dark to light (where blue is seen) is as true of the Benham top as in other colour-forming situations.

The middle of the white half-disc is the flip-over point from ascending to descending brightness in the triplet. Even though I have touched on some processes that relate to this, I have not given a full description of the process.

H) A photograph of the Benham top is colourless

When a still photograph is taken of a spinning Benham top no colour is seen. It is the same as looking at the top when it is not spinning. If the shutter speed is not fast enough, the image will show a blurred black and white top. In this, no contradiction of views is simultaneously placed before the viewer, there is nothing to resolve, and colour does not appear.

However, if the spinning top is filmed, the time difference of the three cone sets comes into play. The transition between dark and light at the arc-ends is reproduced in what the viewer sees. The differing processing rates of the observer's three cone sets are, at any one moment, in conflict. To resolve these different renditions into a single view, a shift to colour vision is called for. The viewer, as a living being that is resourceful, agile and intelligent, is able, through spiritual means (the active mind) to supply the solution. When watching the film, the viewer sees colour on the Benham top.

More succinctly stated: In cases where brightness contradictions are reproduced, the subject's capacity to produce the colours comes into action. In other cases, no colour is seen.

16.10 For curiosity's sake

Is there a position on the midpoint of the black half-disc that compares to the midpoint on the white half-disc?

The position complementary to the midpoint on the white disc should be where the darker sides of the two half-spectra meet – violet and deep red, rather than yellow and light blue. Just as the midpoint of the white half-disc is linked to the formation of green, so the midpoint of the black half-disc should be where magenta appears. Can we conjure this out of the black half-disc? What would be required to produce magenta on a spinning disc? I will not pursue the matter here and leave the question unanswered.

I have not completed the discussion of the Benham top. Diverse features need to be considered to give a full picture of this curious phenomenon. My intent in this little book is to present the unified colour principle and its implications. Since this chapter leads to far more than that, I will put further questions on the Benham top to the side as I believe what has been shown is sufficient to open this line of thinking.

16.11 A parting thought

What holds the flip-over point to the middle of the white half-disc and what sets the dominant half-spectrum of the quadrant? There's more yet to unravel and, for the curious, the journey has only just begun.

17

Q&A On the Benham Top

Q1. During the rotation of the top, what features stimulate image formation?

A1. Image formation is simulated by the passage of the arc-segment.

Q2. How does the passage of the arc-segments activate colour production?

A2. The three cone sets produce contradictions in brightness between their different views. These are resolved, according to the unified colour principle, in the same way that colour resolves the other situations presented in this book.

Q3. At what moment is colour produced?

A3. At the passage of the end of an arc-segment, that is, at the entry into and exit from the arc-segment.

Q4. How are the colours generated?

A4. The three cone sets independently form an image of the passing arc-segment but at different rates. Because of this time difference, image formation is staggered. The transition between light and dark at the ends of the arc-segment does not register at the same instant for each cone set, and this produces contradictory views. In accordance with the unified colour principle, the creative intelligence of the observer generates the specific colour needed to resolve the conflict.

Q5. How do the independent views of the three cone sets become the coloured image we see?

A5. The three views are produced in rapid sequence, too rapid for the conscious mind to discern them individually. They combine in a single view that enters the observer's consciousness.

Q6. How does this process produce the colours we observe?

A6. EXIT from the arc-segment produces colours from the red half-spectrum because the faster responding L cones change to the light ahead of the S cones.

ENTRANCE into an arc-segment produces colours from the blue half-spectrum because the faster responding L cones register darkness (are not registering light) before the slower S cones make the transition.

Q7. How do the different hues in each half-spectrum arise?

A7. Darker hues are produced where the arc-segment touches the black half-disc, either preceding or following. The lighter (whiter) colours occur when a white space separates the arc-segment from the black half-disc. In this situation, both ends of the arc-segment are active, one end is dominant and the other end dilutes (whitens) the dominant hue. The greater the distance from the black half-disc (but not going beyond the midline of the white half-disc), the lighter the hue. It is as if white added to violet/dark blue produces light blue, and white added to red produces orange and yellow. (For yellow and not pink, see Chapter 5.8).

Q8. Where on the spinning top does the coloured image form?

A8. The integration that forms a visible image relates to the passage of the ends of the arc-segments, but the combination of three cone set views takes place in the perceiver's expanded mind. Colour formation is a movement within our psyche and not an event exterior to the mind.

Q9. Why can one not photograph the colour that appears on the spinning top?

A9. A4 answers this question. The different cones form different views, separated in time. A snapshot of the spinning top shows the black and white pattern as it appears when it is stopped or, if shutter speeds are slower, as a blurred image. In both cases, the observer is not faced with the superposition of conflicting views. However, when an observer views a film of the spinning top, the same conflict between different cone sets will arise and colour will appear.

Q10. Why do we see a coloured circle at the radius of the arc-segment?

A10. As the disc turns, the ends of the arc-segment register sequentially on the different retinal cells, producing a circle at the radius of the arc-segment. These reproduce the colour flash all the way around the circle.

Q11. What is the function of the half-black, half-white division of the rotating disc?

A11. The black half-disc without markings creates a pause, enabling a reset for a new cycle of incompatible views.

Q12. Why does the alternation of the white and black halves of the discs not produce colour, yet the same effect at the position of the arc-segments does produce colour?

A12. Because colour formation is linked to image formation. Without the side-by-side relationship of black and white at the arc, no image is formed and no contrasting brightness levels are confronted and integrated. Together with image formation comes the need to resolve conflicting brightness values and it is this that activates the possibility of colour vision.

Q13. Why does the sequence of different coloured circles on the Benham top reverse when the spin direction is reversed?
A13. Because the entry versus the exit sequence of the arc segment is reversed.

Q14. Why, when the direction of spin is reversed, does each coloured circle change to its complementary colour?
A14. When the spin direction of the Benham top is reversed, the contesting brightness levels registered by the three cone sets are reversed.[63]

Q15. How can this description of the production of colour on the Benham top be valid, despite the large difference in the colours observed by different observers?
A15. There are three parts to this answer:
1. The reason for the variation, as discussed in detail in 15.9 and 15.10, has to do with individual differences between the observers in both visual acuity and rate of processing. These factors come to the fore because, in the case of the Benham top, the time to form the image is exceedingly short in comparison with other situations where colour is observed.
2. The different colour observations are classified in terms of the half-spectrum observed (blue or red). In grouping observations in this way, small variations are less important and we recognise more easily the overall pattern in the observations.
3. The limited consensus between different observers relates to point 1, but there is still a significant degree of consensus. Even if only 30 per cent of viewers agree, this shows a common

---

[63] A prerequisite to understanding the change in colour produced by the reversal of the direction of spin is a recognition of the colour sequences in the two half-spectra and how they are produced (Chapter 5). The contesting brightness levels of the three cone sets that produce one half-spectrum is the reverse of those producing the other half-spectrum.

trend. This suggests that even with the wide variation between observers (point 1), there are underlying principles of vision that create an agreement. The presence of consensus suggests the process is not random but that it has an underlying structure.

About the Author

Peter Moddel was born in Ireland and currently lives in Gruyere, Switzerland. His studies in various disciplines (philosophy, literature, physics, pedagogy), the experience of living both in the West and the East, and periods of personal retreat fostered his reflections in philosophy, science and linguistics. Astronomy has been central in his activities and building telescopes expanded into an interest in vision and in colour theory. Understanding colour formation became a subject of personal research. Other passionate pursuits include hiking and music. He co-leads monthly café-philo encounters that aim to promote agility of thought and freedom from entrenched viewpoints.

Index

Page numbers followed by *f* indicate figures. Page references including an n indicate notes.

Q

R

9 782970 096726